THE WORLD
OF DINOSAURS

The University of Chicago Press, Chicago 60637
Preface © 2019 American Museum of Natural History
Text © 2019 Mark Norell
Design © 2019 Andre Deutsch

Published 2019
Printed in Dubai

28 27 26 25 24 23 22 21 20 19 1 2 3 4 5

ISBN-13: 978-0-226-62272-9 (cloth)
ISBN-13: 978-0-226-62286-6 (e-book)
DOI: https://doi.org/10.7208/chicago/9780226622866.001.0001

LCCN: 2018041665

THE WORLD
OF DINOSAURS

An Illustrated Tour

MARK A. NORELL

The University of Chicago Press

CONTENTS

THE AMERICAN MUSEUM OF NATURAL HISTORY

The American Museum of Natural History, founded in 1869, is one of the world's preeminent scientific, educational, and cultural institutions.

The Museum encompasses 45 permanent exhibition halls, including those in the Rose Center for Earth and Space and the Hayden Planetarium, as well as galleries for temporary exhibitions. It is home to the Theodore Roosevelt Memorial, New York State's official memorial to its 33rd governor and the nation's 26th president, and a tribute to Roosevelt's enduring legacy of conservation. The Museum's five active research divisions and three cross-disciplinary centers support approximately 200 scientists, whose work draws on a world-class permanent collection of more than 34 million specimens and artifacts, as well as specialized collections for frozen tissue and genomic and astrophysical data, and one of the largest natural history libraries in the world. Through its Richard Gilder Graduate School, it is the only American museum authorized to grant the Ph.D. degree. Beginning in 2015, the Richard Gilder Graduate School also began granting the Master of Arts in Teaching (MAT) degree, the only such freestanding museum program. Annual visitation has grown to approximately 5 million, and the Museum's exhibitions and Space Shows are seen by millions more in venues on six continents. The Museum's website, mobile apps, and MOOCs (massive open online courses) extend its scientific research and collections, exhibitions, and educational programs to additional audiences around the globe. Visit amnh.org for more information.

PREFACE

Dinosaurs occupy a special place in our culture. They are not only the subject of exciting and important discovery in science; they are extraordinarily alluring to masses of people, young and old. The American Museum of Natural History in New York, where the author of this book, the well-known dinosaur paleontologist Dr Mark Norell, works, bears testament to this fact, as it annually greets over 5 million visitors, many of whom, as Mark points out, simply call it the "dinosaur museum".

Why are dinosaurs so popular? This is perhaps the most frequent question asked of paleontologists like me, one almost certain to arise on numerous occasions, whether at a press event, an educational conference, a holiday gathering of extended family, or a cocktail party. In his introduction, Mark ventures an answer: these spectacular creatures inspire and, at the same time, challenge our imagination. They are (often) gigantic and extreme in ways unpredicted by our experience in the living world. And their fossil remains tell us that an ancient world so alien to our own is not a fantasy, but a reality.

What is that reality as we know it now? How much breaking news is there on the dinosaur research front? As this book shows, a lot. For one thing, the recognition that birds are actually living dinosaurs – the one branch that somehow survived the so-called dinosaur extinction event of 66 million years ago – is a triumph of modern paleontology. What was once so controversial and even discredited is now so well-bolstered by scientific evidence that it is overwhelmingly endorsed by scientists, an endorsement that has not escaped broad public awareness. Yet this breakthrough is only the beginning. Powerful new insights about dinosaurs have come from many directions – recent discoveries of exquisitely preserved and extraordinary fossils, use of the latest in imaging and analytical tools, and of course our acute knowledge of the biology of living dinosaurs, the birds. We now know much about ancient dinosaurs that was only a matter of speculation or fantasy a few years ago – how fast dinosaurs grew, how they reproduced and cared for their young, and even what colour they were.

All this activity and discovery has only served to enhance the popular appeal of dinosaurs. Something so captivating is bound to give rise to legend and myth. Dinosaurs have attracted a good deal of misconception and folklore, notions that are merely distortions of what we really know and don't know about them. One does not have to be a scientist to write a good book about dinosaurs, one that draws the line between fact and fantasy. But scientists, like the author, who can imbue their writing with their expertise and experience as a dinosaur hunter and researcher, offer something special. Indeed, Mark has a right to claim that many of the extraordinary discoveries he describes herein were made by him and his colleagues. The result is an unusually authentic and vivid introduction to dinosaur science.

Michael J. Novacek
Senior Vice President and Provost of Science; Curator, Division of Paleontology American Museum of Natural History

INTRODUCTION

ALMOST EVERYONE ON THE PLANET KNOWS OR THINKS
THEY KNOW WHAT A DINOSAUR IS. THEY ARE EVERYWHERE.
THEY PERMEATE POPULAR CULTURE THROUGH ADVERTISING,
CHILDREN'S TELEVISION AND AS A METAPHOR FOR "OLD",
"EXTINCT", "OBSOLETE", "DUMB" OR "UGLY".

They have been the subjects of film nearly since the medium was invented. As an animated feature, *Gertie the Dinosaur* appeared in 1914, just a few years after the first animated films. In the ensuing years real star power was associated with dinosaurs. The opening of *Bringing up Baby* in 1938, starring Katharine Hepburn and Cary Grant, featured Grant as a goofy paleontologist in search of an "intercostal" bone to finish his mount of a *Brontosaurus*. More star power quickly followed. In 1940 Disney studios and the composer Igor Stravinsky collaborated to craft the epic *Fantasia*. This animated film featured dancing dinosaurs strutting to Stravinsky's *The Rite of Spring*. My predecessor at the American Museum of Natural History, Barnum Brown (the man who discovered *Tyrannosaurus rex*), consulted on this project. Brown's efforts eventually led to a major attraction at the 1964 New York World's Fair and a permanent attraction at Disneyland. Other more recent dinosaur movies have run the gamut from fairly accurate scientific documentaries to films such as soft-core porn *Dinosaur Valley Girls*. Believe it or not the online fiction business, if you look at Amazon, even has a thriving erotic dinosaur fantasy section. Search at your own risk – although the titles are hilarious.

The connection between dinosaur science and Hollywood has never been so apparent as in the *Jurassic Park* franchise. The first *Jurassic Park* movie appeared in 1993. Directed by Steven Spielberg, based on a Michael Crichton novel, and with an all-star cast, it has made well over a billion dollars at the time of writing. When it was released, this was the most realistic portrayal of dinosaurs that the movie-going public had ever seen. Using a combination of, by today's standards, fairly primitive animation and puppets, dinosaurs were brought to life. Since then it has morphed into a franchise of four feature films with another in production. Millions of dollars have been spent and millions of people have seen these films, although the underlying premise is based on some fairly dubious science and really misrepresents what dinosaur paleontologists do. Like Disney's *Fantasia* before it, *Jurassic Park* has also led to theme park attractions.

The pop-glitz of Hollywood is not the only barometer of the public's fascination with these ancient reptiles. Dinosaur displays and exhibits pack them in. Millions of people attend museums, universities, event centres, science centres, even shopping malls where dinosaurs are on show. Interest and fascination in dinosaurs is a universal in an ever-more divided international community. My own museum, the American Museum of Natural History in New York, receives over 5 million visitors a year, and the dinosaur halls are by far the most popular exhibits. So

> The David H. Koch Hall of
Saurischian Dinosaurs at the
American Museum of Natural
History. This is one of six
fossil vertebrate halls at
this venerable Museum.

much so, that the Museum is popularly referred to as the "Dinosaur Museum".

There is a common conception that all of these films (especially *Jurassic Park*) have created global interest in dinosaurs. Surely, from a historical perspective this is not the case. Dinosaur reconstructions crafted by Benjamin Waterhouse Hawkins and shown at the Crystal Palace in 1854, when it was relocated from Hyde Park to South London after the Great Exhibition, were a huge visitor attraction, and throngs of crowds attended the unveiling of successive dinosaur exhibits at the American Museum of Natural History in the early twentieth century.

So why all the interest? As a professional myself, it is a question that I am constantly asked. There has been much written, and much commented on. But to me it simply has to do with our imaginations. Everyone the world over knows about animals in some sense. They can be the pigeons we see in the park, the cows and pigs we eat, exotic animals at a zoo, or even more exotic ones on television. However, none of these could prepare us for dinosaurs – many of them spectacular beyond spectacular. They are the superstars of the ancient world. Gigantic animals, spiked and horned, fanged, fluffy, fearsome and fast. Just seeing their skeletons in museum displays brings smiles and amazement (or occasionally, tears, in the case of young children). To look at a skeleton of one of these animals stimulates our curiosity. We can't directly observe the look, behaviour, or diets of these extinct beasts the way we can a New York City pigeon, but we can use our imaginations. And that is much of what people do. They bring the museum bones to life, each in their own way.

I have so far ignored the science behind it all. The way that dinosaur paleontologists, for the

∧ A still from the 1938 film *Bringing up Baby* starring Cary Grant and Katharine Hepburn.

< Gertie appearing in the eponymous 1914 film *Gertie the Dinosaur*, one of the first animated films.

> Dr Alan Grant (portrayed by
Sam Neil) confronts a very
vile pack of Velociraptors
in the 2001 film *Jurassic Park
III*.

∨ Benjamin Waterhouse Hawkins'
life-size sculptures of
Iguanodon in London's Crystal
Palace Park.

most part, look at this is quite different. All of us in the business come at it with our own style, interests, experiences and perspectives. The scientific questions that are entertained run the gamut from understanding how these animals were related to one another, what they ate, how they behaved, where they lived, even what their brains were capable of, what colours they were or what they could see. These are the same areas of study that modern biologists apply to extant animals. Basically, we are biologists who work on fossils.

One of the things that makes dinosaur study so valuable is its importance to science education. Media saturation and underlying interest have resulted in people probably knowing more about dinosaurs than most other groups of animals, living or extinct. They know that an asteroid may have hit the planet just over 65 million years ago. They are familiar

with all sorts of dinosaur names. They may have an opinion on whether *Tyrannosaurus* was a scavenger or predator. The list goes on, and it is our responsibility as dinosaur scientists to leverage this interest into educating people about a whole range of subjects that without dinosaurs might seem dry and uninteresting – genetics, earth history, mathematics, functional morphology and computer science. As a professional, it is heartening that the things

I pull out of the ground can have such far-ranging effects.

Finally, there is so much dinosaur paleontology, movies, advertisements, and souvenirs out there it is difficult to define a reality line. There is a lot of confusion and misinformation and it is hard to sort out real discovery from tabloid news, overzealous screenwriters, shoddy science, uninformed bloggers, etc. Let's start to make sense of this.

∧ Even in the early 1960s dinosaurs were popular enough to be a major public attraction. These were floated around Manhattan Island on a barge before their installation at the 1964 World's Fair in New York City.

ABOUT THIS BOOK

BY MANY CALCULATIONS THE AMERICAN MUSEUM OF NATURAL HISTORY HAS THE MOST IMPORTANT COLLECTION OF DINOSAURS IN THE WORLD. IT MIGHT NOT BE THE LARGEST (CALCULATED ON THE BASIS OF THE NUMBER OF ACTUAL BONES), BUT IT IS THE MOST DIVERSE.

It is the legacy of over 125 years of dinosaur collecting by successive paleontologists at the Museum, including Henry Fairfield Osborn, Walter Granger, Barnum Brown, Edwin Colbert, and a plethora of technicians and students.

The collection was amassed from excavations around the planet, giving the Museum collection more of a synoptic and cosmopolitan character than many others. It was assembled in the first Golden Age of dinosaur collecting (roughly the last decade of the nineteenth century and the first decade of the twentieth century), a world far different to today. Before the Second World War it was much easier to collect dinosaur fossils in developing countries and retain them in your own collections. Because there was no commercial value for fossils, private land was open territory, usually with the generous hospitality of the land owner. For the most part, and certainly in the case of the American Museum of Natural History, even though regulations were not as severe, permissions were still appropriate and required. This was not Indiana Jones or Lara Croft territory. For instance, the Museum trips to Mongolia's Gobi Desert were under the auspices of the Mongolian government whose permits the Museum's archives still retain. Today most countries do not allow the permanent export of their dinosaur treasures. Yet this has not deterred the exploration of classic, far-flung or new fossil-producing areas – instead, we are in a new, more internationalist, Golden Age of dinosaur collecting.

The Museum's dinosaur displays have been organized in their current incarnation since 1995. Groundbreaking at the time, the dinosaurs and other fossil vertebrates were laid out in a plan with their living relatives which reflected their relationships to one another and also documented the evidence supporting these hypotheses. Previous displays were walks through time. Fortunately, the phylogenetic structure in our halls has remained relatively stable since the galleries opened. To tell this story in a complete and informative way required us to go afield from our permanent collections and add casts of specimens to fill holes in the story. These include casts of virtually all of the important specimens that Museum expeditions have collected in Mongolia, where I have been excavating since 1990 (Mongolia no longer permits permanent export of its specimens).

The last couple of decades has resulted in more dinosaur research being undertaken. Dinosaurs are being found, described and studied at a faster rate than ever before. And, in addition to augmenting the number of species being found, spectacular specimens abound. Dinosaurs with feathers are now common, while dinosaurs sitting on their nests brooding their eggs, dinosaur embryos, and other bizarre and unexpected discoveries have been uncovered across the globe.

> *Triceratops* greets visitors to the David H. Koch Hall of Ornithischian Dinosaurs.

e the horns and frills
ceratopsians for?

erly thought that ceratopsians used their
against predators like *tyrannosaurus rex*,
ge frill served mainly as an attachment site
which presumably would have resulted in a

Triceratops, a horned dinosaur

The way in which we study dinosaurs has changed as well. Instead of a magnifying glass and a brush, we are more likely to use synchrotron radiation, mass spectrometers, and some of the most advanced computers in the world. My colleagues are just as likely to be engineers, molecular biologists, and computer scientists as they are traditional paleontologists.

This book is about all of this. It is not a synoptic account of dinosaurs around the world. It is the story of contemporary multidisciplinary dinosaur research and all

of the new recently discovered animals told through the iconic dinosaurs on display at the American Museum of Natural History. As indicated above, the order of the dinosaur specimens that are featured in this book mimics their placement in the two dinosaur halls at the Museum. Not all of the dinosaurs on display at the Museum are covered here, but the major ones are, and most are mentioned. It is through the lens of these species that we examine history, science and the appeal of these incredible animals.

∧ A young Barnum Brown (left) and museum director Henry Fairfield Osborn at Como Bluff, Wyoming, USA in 1897 excavating a specimen of *Diplodocus*.

> A reconstruction of the Early Cretaceous tyrannosaur *Yutyrannus*. The fluffy feather coating is known from fossils of this specimen which were collected in Northeast China.

WHAT ARE DINOSAURS?

MOST PEOPLE KNOW, OR THINK THEY KNOW, WHAT A DINOSAUR IS. THEIR DEFINITION USUALLY GOES ALONG THE LINES OF "HUGE, SCALY, EXTINCT, OBSOLETE, ETC." THIS COULD NOT BE FURTHER FROM THE TRUTH. LET'S LOOK AT WHAT DINOSAURS REALLY ARE.

D inosaurs are a group of reptiles that first appear in the fossil record about 235 million years ago during the latest part of the Triassic Period (~252 to ~201 million years ago). Dinosaurs are a group that paleontologists deem to be "monophyletic", which means that all members of this group sprang from a single common ancestor.

There are many animals that look like dinosaurs to the uninitiated. These include marine lizards – the mosasaurs – and very un-crocodile-looking primitive crocodiles – the crocodilians – some of which were totally terrestrial, bipedal and in some cases lacked teeth.

Dinosaurs belong to the group Dinosauria. This term was first used in 1842 by the English comparative anatomist Richard Owen (1804– 1892). It is derived from the Greek meaning "terrible reptile". Owen didn't propose the word "terrible" in the negative sense meaning unpleasant; rather, he used it as the common contemporary British adjective meaning "very great". Undoubtedly this was in reference to the tremendous size of the fossil dinosaurs known at the time, which were first found in the English countryside.

Soon after Owen coined the term, many new fossils were discovered in Europe and North America, and it became apparent that dinosaurs were a very successful group that spanned an extensive stretch of geological time. Before too long, so many discoveries had been made that it was possible to place them in what is called a phylogeny, which is the specific term for a genealogy or family tree.

These discoveries allowed the Victorian paleontologist Harry Seeley (1839–1909) to propose in 1888 that there were two great groups of dinosaurs: the Saurischia (reptile-hipped dinosaurs) and the Ornithischia (bird-hipped dinosaurs). Seeley divided dinosaurs, as the names suggest, into these groups on the basis of their hip structure (see p.20). From his perspective, the hips of dinosaurs such as *Iguanodon* and *Hylaeosaurus* (both species that were well-known in British collections) looked like those of modern birds, where a large part of one of the pelvic bones (the pubis) points towards the tail of the animal. In the more typical saurischians the pubis pointed forward – as it does in not only most other reptiles, but also for the

⌐ Thomas Huxley was a staunch defender of Charles Darwin's views on evolution. It was Huxley who first recognized the relationship between birds and extinct dinosaurs.

∧ The venerable Victorian biologist Richard Owen. Although never accepting Darwinian evolution, Owen coined the word Dinosauria.

> It would take a lot of imagination to conjure up an animal as bizarre as *Triceratops*. It would be hardly believable if we didn't have actual fossils.

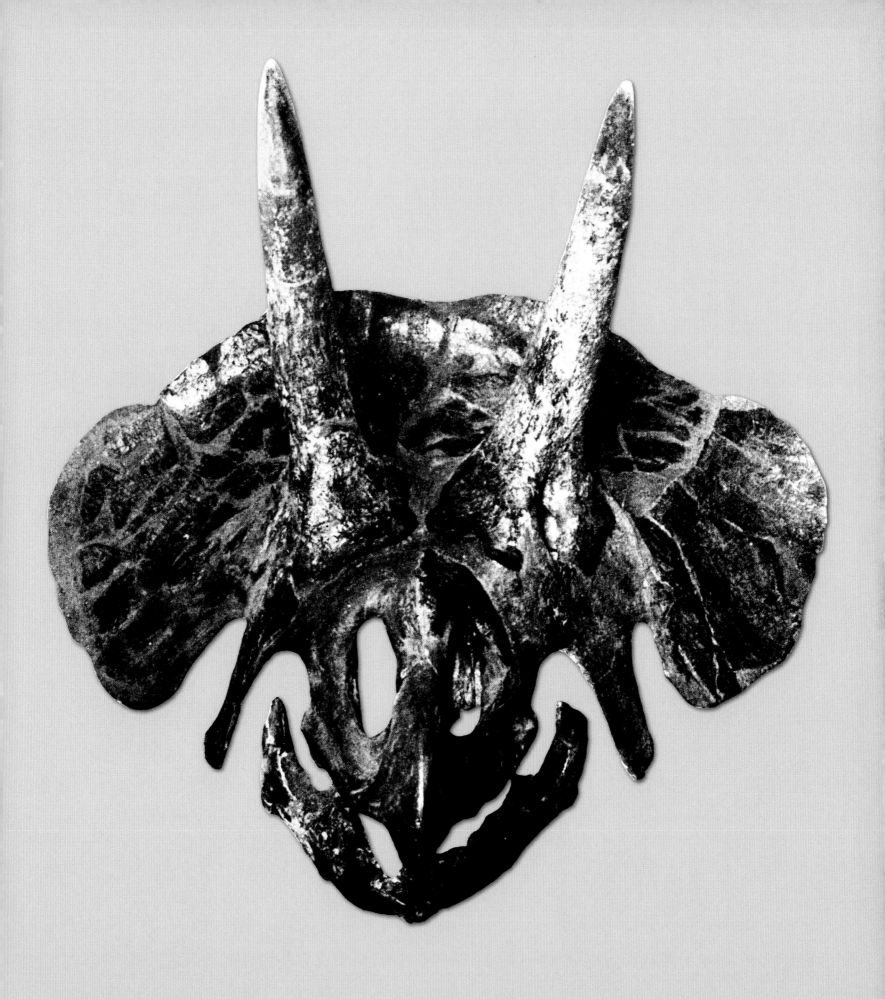

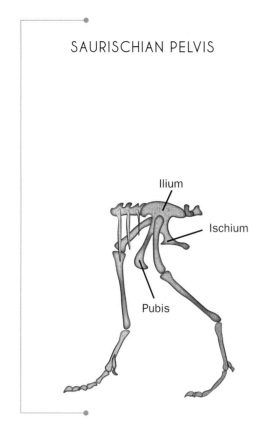

SAURISCHIAN PELVIS

Ilium

Ischium

Pubis

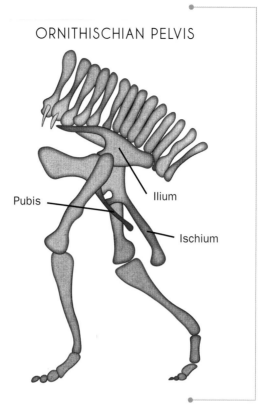

ORNITHISCHIAN PELVIS

Pubis

Ilium

Ischium

< The difference in the
orientation of the hip bones
is the traditional defining
characteristic between the two
major groups of dinosaurs. In
Ornithischia (right) the pubis
points backward. In Saurischia
the pubis points slightly
forward.

majority of other tetrapods (the group that includes all land-living vertebrates).

Unfortunately, Seeley's proposition confused things immensely in the few decades following. Beginning in the 1860s, as will be discussed later, the idea was put forward that birds evolved from dinosaurs. Thomas Huxley (1825–1895) was the first to recognize this based on his studies of *Megalosaurus* and the London specimen of the protobird *Archaeopteryx*. The root of the confusion is that the dinosaurs from which Huxley postulated that birds evolved were not the bird-hipped ornithischians, but the reptile-hipped saurischians. It turned out that Seeley had made an interpretive error in describing the ornithischian pelvis. This error stems from the very highly derived nature of the ornithischian pelvis.

The pelvis of tetrapods (four-limbed animals) is composed of three paired elements: the ischium, the ilium and the pubis. These three paired elements form a platform to support the rear appendages – the legs. The three elements meet around what is called the acetabulum, which is a shallow socket where the head of the femur (the thigh bone) attaches to the torso in a ball and socket joint. In most animals, all of the pairs of bones meet on the midline. The ilia meet with fused backbone segments (vertebrae)

to tightly anchor the hind limbs in what is called a sacrum to the torso. As indicated above, the pubes generally project downward and forward, mirroring the ischia which project backwards and downwards.

Living birds are an exception. In these animals neither the paired ischia nor the pubes converge and meet each other along the midline. Furthermore, the pubis, instead of being directed forwards, as it is in most animals and in most non-avian saurischian dinosaurs, points backwards like Seeley thought the pubis of ornithischian dinosaurs did.

If we take a closer look at the pelvis of ornithischian dinosaurs, we can see how Seeley made his error. In the "bird-hipped" dinosaurs the pubis has two components. It has a forward pointing process and another process that projects backwards. Modern thinking is that the backward pointing process is what we call a "neomorph" – a novel characteristic. Yet many paleontologists consider a small element, the propubis in ornithischians, to be the element that actually has an evolutionary history that is tied to the ancestral pubis. This is called a homology.

Homology is an important – probably the most important – theoretical concept for the study of evolution. Literally, homology means

⅂ Mosasaurs were large
seagoing lizards, not related
to dinosaurs. They were top
predators in shallow Late
Cretaceous seas.

> *Effigia* was a specialized
bipedal crocodile-line
archosaur — not a dinosaur. It
is about 2m (6½ft) long and is
found in the same sediments as
Coelophysis (see p.56)

"sameness". However, it refers to a particular kind of sameness, a sameness that is inherently tied to evolutionary history. Hence it provides evidence for the inter-relatedness of organisms. This is best explained through example.

All vertebrate animals have backbones. The explanation for this is that the backbone evolved once and was therefore present in all of the descendants of the common ancestor of vertebrates. In science the simplest answer is always considered the best, as it requires the least number of special explanations, or assumptions. For instance, if we said that the backbone evolved twice, that requires two steps as opposed to one in a single origin. Within vertebrates a subgroup of animals has four limbs – the tetrapods. This group includes familiar animals such as amphibians (frogs, salamanders and their kin) and another more advanced group (characterized by their unique attribute – an egg that can be laid on land) called amniotes. The amniotes include the lineage that incorporates turtles, lizards and snakes, crocodiles, many extinct forms and, most importantly here, dinosaurs (which includes birds). The other lineage of amniotes includes all of the animals leading up to modern mammals. Many of the mammal lineage animals, especially early representatives, such as the fin-backed pelycosaurs, or the dim lumbering dicynodonts, are often confused with dinosaurs. Since these animals are not descended from the first dinosaur and are more closely related to mammals, they are not classed as dinosaurs.

Among living animals, the closest relatives to dinosaurs are crocodiles. They have several characteristics (homologies) in common. These include many features of the skeleton as well as characters from their DNA. Perhaps, surprisingly, they also share some morpho-physiological characters with dinosaurs, such as an advanced one-way lung ventilation system, and even behavioural characteristics like nest guarding and possibly cooperative hunting. But the early history of crocodilians was very different; to the untrained eye, many of these animals did not look crocodile-like at all.

The group that includes dinosaurs (including birds) and crocodilians and their relatives is called Archosauria. Archosauria means "ruling reptiles". The archosaurs are divided into two monophyletic groups, the bird-line archosaurs

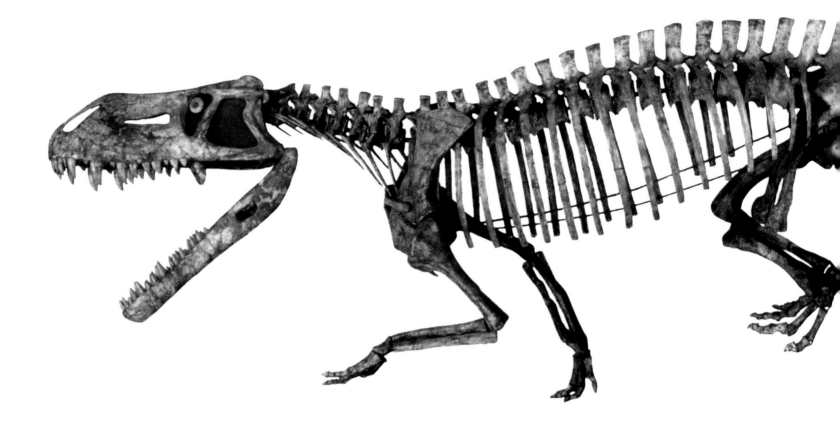

and the crocodile-line archosaurs. The bird-line archosaurs are familiar. These include living dinosaurs (the birds) as well as other animals, the non-avian dinosaurs and also the non-dinosaurian flying reptiles – pterosaurs. Less diverse, and largely unfamiliar are the crocodile-line archosaurs. Our understanding of this group is shackled by a limited living diversity (23 or so species) restricted to a single circumequatorial habitat and a paucity of fossil remains. But what has been established is that crocodile-line archosaurs were diverse in both body form and species richness. Many of these animals were fully terrestrial. Some were large carnivores, like *Prestosuchus*; others, like *Zarasuchus*, were small ecological specialists with tiny teeth, gracile limbs, and spiky body armour. Others such as *Dakosaurus* were pelagic seagoing giants, up to 5m (16½ft) in length.

It is necessary here to introduce the concept of disparity. When we think of the success of a group of organisms in evolutionary time there are a couple of ways to measure it. The one that is most familiar to both scientists and non-scientists is abundance. This is simply a measure of the number of species through time. Often this is portrayed as a simple

bivariate plot of the number of species found in the fossil record plotted against absolute time. This approach is fraught with difficulties and error, as a very large component of this analysis is not the actual number of species, but how many fossils have been found. This means that during time periods for which we lack sediments that contain fossils, the world looks like a very species-poor place. While some analytic methods allow us to correct for this, it is still an inadequate method of measuring diversity.

Rather than measuring the abundance of species, disparity measures how different organisms are from one another. As an example, there are many different species of birds called warblers. They are incredibly diverse, with up to 50 species in the Americas alone, but they all look very similar – so much so that hardcore birdwatchers have difficulty telling them apart. A much less diverse group, ratites, for example, contain only about 12 living species and a number of fossils. These range from the diminutive Kiwi to the recently extinct giant elephant birds (*Aepyornis*). So which group is more diverse, the speciose warblers or the morphologically different

ratites? That is a good question. While warblers are more diverse, ratites are certainly more different from one another. The last few years have seen the development of analytical tools that can measure disparity. Unsurprisingly, disparity and diversity are often not highly correlated over time. However, many – if not most – scientists would agree that both disparity and diversity need to be examined to understand the evolutionary dynamics of life through time.

While crocodile-line archosaurs never achieved the species diversity or the morphologic disparity of their closest relative the bird-line archosaurs, they were still much more disparate than their living diversity reflects. The most primitive members of the group look so similar to early bird-line archosaurs that it is difficult to tell them apart. They were bipedal, perhaps feathered, carnivorous and shared many characteristics with dinosaurs and birds.

Another issue that adds to the conundrum of defining dinosaurs is the fact that so many animals found in recent years are very closely related to dinosaurs, but do not lie within the groups Ornithischia or Saurischia.

Although their pelves are of the saurischian type, they lack several characteristics that are present in both saurischians and ornithischians. Collectively we call these animals "dinosauromorphs", and they are more closely related to dinosaurs than pterosaurs are. Because these animals (such as *Dromomeron*, a Late Triassic dinosauromorph from the Ghost Ranch locality in New Mexico) have many features of dinosaurs not seen in crocodile-line archosaurs, they somewhat blur the line between dinosaur and non-dinosaur, leaving little that remains to define what a dinosaur actually is except for some technical anatomical minutiae, and an important character explained below. One of the groups that has been proposed to be the closest relative of dinosaurs is the Silesauridae. These were poorly known diminutive animals, which belong to a group called the Dinosauriformes. They were quadrupedal herbivores, but they share a common characteristic with all dinosaurs – a hole in the hip socket. The hole in dinosauriformes is small but it is there as a small slit. In dinosaurs it is a large opening and this is one of the best characters by which to recognize a true dinosaur.

Although there are several characteristics that define dinosaurs, when all the data is collected there is no singular attribute, except for perhaps the fully perforated hip socket, that we can use to diagnose the group. Part of the reason for this is that direct character evidence for dinosaur monophyly is so scarce and that there have been so many close dinosaur relatives, and so many primitive crocodiles, found in the last several years. Nevertheless, all dinosaurs are still considered to be descended from a common ancestor and are therefore more closely related to each other than any is to another vertebrate.

Seeley's view that there were two kinds of dinosaurs has been a stable and orthodox view for almost 150 years and is the framework within which generations of paleontologists have interpreted their results. This orthodoxy has recently been challenged. In 2017 a research group based in the UK concatenated a very large data set and reanalyzed this problem. They concluded that, instead of Ornithischia and Saurischia, ornithischians are more closely related to a subgroup of conventional Saurischia, the theropods, which includes birds to the exclusion of the Sauropod dinosaurs.

Sauropods have always been considered to be a member of Saurischia since Seeley.

Their conclusion was based on an analysis of a very large data set of characters. Sadly, some of this data set is compromised by the fact that they did not observe many of the specimens in person, nor did they include the complete retinue of specimens. Quickly there were replies that supported the old orthodoxy. But not by much. Because of the plethora of new discoveries, coupled with all the work in this area by other young scientists, this is certainly an area to watch. But, as of this writing, the plurality of dinosaur paleontologists are still working under the paradigm of Seeley.

∧ Some primitive crocodile-line archosaurs look distinctly un-crocodilian. This is *Prestosuchus*, from Middle Triassic (~237 million-year-old) beds in Brazil. It was a large (nearly 3m/9¾ft) cursorial predator.

A HISTORY OF DINOSAUR DISCOVERY

DINOSAUR FOSSILS HAVE UNDOUBTEDLY BEEN FOUND AND COLLECTED FOR MILLENNIA. AFTER ALL, THEY ARE RATHER OBVIOUS, AND WE HUMANS ARE CURIOUS BY NATURE. THEIR DISCOVERY IN A SCIENTIFIC CONTEXT, HOWEVER, BEGINS IN WESTERN EUROPE AND THE SCIENTIFIC ENLIGHTENMENT THAT OCCURRED THERE.

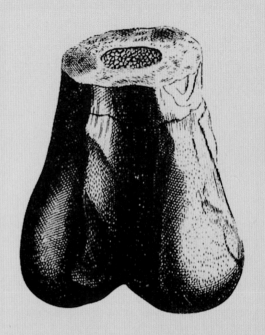

A wealth of new ideas was developing as the burgeoning scientific community began to explain the world of biology, physics, medicine, geology, etc., with the tools of science as opposed to theology, superstition and belief. The first record we have of what is definitively a dinosaur comes from 1676 when Robert Plot (1640–1696), then head of Oxford's Ashmolean Museum, published a short note on a bone found in Oxfordshire that had been given to him. It is, of course, probable that the Chinese found similar specimens long before, as records going back to at least the Jin Dynasty (AD 1125–1234)

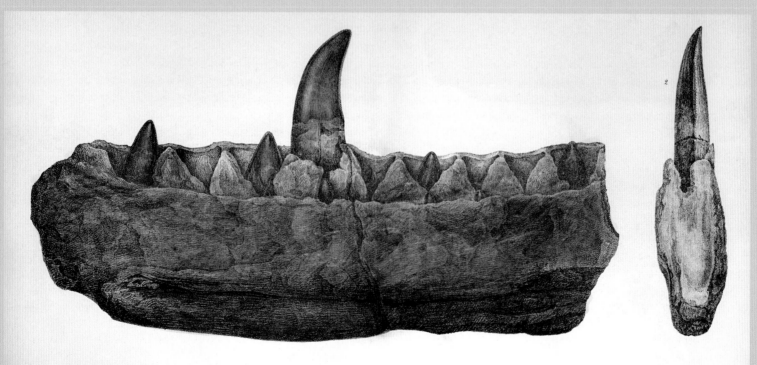

ANTERIOR EXTREMITY OF THE RIGHT LOWER JAW OF THE MEGALOSAURUS.
FROM STONESFIELD NEAR OXFORD.

Scale of Inches

and perhaps the Han (206 BC – AD 220) report the presence of "Dragon Bones" in southern China. Dragon bones (*long gu*) were then (as today) a common ingredient in the Chinese apothecary, used to soothe the spirit, to rid one of ghosts, and to calm the nerves, as well as to cure everyday ailments such as sores and alleviate menstrual symptoms. Most Chinese vertebrate fossils destined for the apothecary market in more recent times come from Shanxi Provence and are the fossil remains of mammals rather than dinosaurs – but certainly dinosaurs must have come into the mix. The Jin Dynasty bones were thought to come from Sichuan, which today is renowned for both dinosaur and mammal fossils.

Plot recognized that the bone he received in 1676 was the end of the femur (the upper leg bone) which formed part of the knee of a very large animal. Variously he considered it to be the remains of a giant antediluvian human

or a Roman war elephant. Comically, nearly 100 years later in 1763, English physician and author Richard Brookes (1721–1763) described this bone and labelled it "*Scrotum humanum*", in reference to its distinctive shape. Sadly, the element has been lost in the intervening years, but from contemporary published illustrations, it is obvious that it is the distal end of the femur (the upper part of the knee) of *Megalosaurus*, a now common saurischian carnivorous dinosaur.

Within the modern scientific framework, the first fossils recognizable as dinosaurs were collected by English gentry in the southern part of the country. Remains were first reported by William Buckland (1784–1856), an Oxford professor. He recognized these as belonging to a group of large carnivorous reptiles and coined the name *Megalosaurus* (large lizard) in 1824. Other dinosaur fossil discoveries soon followed. Mary Ann Mantell (*c*.1795–*c*.1855) had found fossil remains of a large reptile

∧ Chinese farmers excavating "dragon bones" in Central China in the 1920s. Although these were the bones of fossil mammals, they were still considered to be an important part of the Chinese apothecary.

⌐ Although now lost, this is the first dinosaur bone to be described and illustrated. This illustration of the lower end of the femur (thigh) in *Megalosaurus* was figured by Robert Plot in 1676 and later labelled by Richard Brookes as "*Scrotum humanum*".

< An illustration of a *Megalosaurus* jaw from William Buckland's treatise "Notice on the *Megalosaurus* or great fossil lizard of Stonesfield" in 1824. This is one of the first illustrated dinosaur fossils.

on an excursion to the English countryside in 1822. On the inspection of her husband Gideon Mantell (1790–1852), these turned out to belong to a herbivorous contemporary of *Megalosaurus*. Because of the tooth's common characteristics with the New World herbivorous lizard *Iguana* (albeit larger in size), he named it *Iguanodon* in 1825, one year after Buckland announced his discovery.

From there, fossil finds started to pile up around the world, including the serendipitous discovery of 38 *Iguanodon* skeletons in a single coal mine at a depth of over 300m (984ft) in Belgium in 1878. Such accidental discoveries became more commonplace, but the era of organized scientific excavation had begun, primarily in North America. It was there that the great dinosaur collections were made in the American West, beginning with westward

expansion after America's Civil War. For a long time these were the most important dinosaur collections in the world. In recent years, while not eclipsed by new discoveries, the notable collections in Argentina, South Africa and especially China have given us a much better picture of these remarkable animals.

Today dinosaur discovery occurs at a pace unmatched during the entire history of paleontology. Paleontology has truly gone global and discoveries from outside North America and Europe outnumber those from the traditional continents. To non-scientists, it may seem that having lots of new data is great for giving answers to old problems. To those of us who practise, this is a double-edged sword, as some of the new discoveries, which you will read about in this book, open more new questions than they answer old ones.

⌐ Joseph Leidy was an important American paleontologist. In 1858 he described *Hadrosaurus foulki*, at the time the most complete dinosaur skeleton yet recovered.

∧ Mary Ann Mantell found the first specimens of *Iguanodon*.

⌐ By the last quarter of the nineteenth century reconstructions of extinct fossil reptiles became all the rage. Here *Iguanodon* and *Megalosaurus* are depicted as slovenly reptilian slobs.

> Gideon Mantell's 1834 drawing of *Iguanodon*. He misinterpreted the large thumb spike as a rhinoceros-like horn and for some reason depicted the animal, which was over 10m (32¾ft) long, climbing on a tree branch.

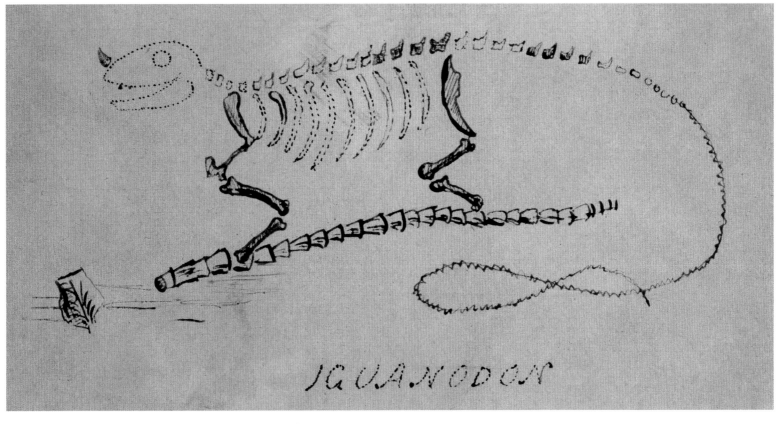

IGUANODON

GEOLOGICAL TIMESCALE

EARLY ON IN THE HISTORY OF GEOLOGY, PEOPLE RECOGNIZED FUNDAMENTAL PRINCIPLES ABOUT HOW SEDIMENTARY ROCKS WERE DEPOSITED ON EARTH. SCHOLARS BEGAN TO CODIFY THESE IN THE SEVENTEENTH CENTURY.

O ne of the first was the rule of superposition. This was proposed by Nicolas Steno (1638–1686), the Danish cleric later proclaimed a saint. Steno was a true polymath, who in 1669 formally proposed what to us is fairly obvious: superposition simply means that in rock strata, the sedimentary layers nearer the top are younger than the ones lower down.

For example, if you trek down into the Grand Canyon, the rocks on the banks of the Colorado River are older than the rocks at the lookout on the rim. As you go deeper into the Earth, you go deeper in time. For paleontologists this is significant, because fossils that are found in rocks in lower strata are older than fossils found in rock strata above them. Although the rock layers may be twisted and folded by Earth processes, this general realization quickly allowed early paleontologists to recognize that the older rocks contained fossils of more primitive life forms than fossils contained in overlying rocks.

Fossils are the only record that we have of ancient life on the planet. Fossils are any trace of the organisms that lived in the past. They range from chemical signatures, to footprints, to actual fossil bones. In this book we will be primarily concerned with what are called replacement fossils. These are formed (as their name suggests) by the replacement of the original biological material with some other mineral. This replacement can be very exact, down to microscopic levels, which allows minute structures, such as individual bone cells,

to be preserved and allows us to address an entire retinue of paleobiological questions. As will be explained later, these include analyses of growth and longevity.

Another kind of fossil is a trace fossil. The most familiar type of trace fossils are trackways. Dinosaur tracks are very common and we can learn a lot about these animals from detailed examination of their tracks. Animal speeds, the soft tissue covering of the feet, even the fact that they were social animals, have all been deduced from tracks. Finally, it should be remembered that fossilization of any type is extraordinarily rare, and we have fossil remains for an infinitesimally small number of the actual number and kinds of plants and animals that have ever lived.

Steno's idea of superposition is one of the two major geological principles that are important to paleontologists. The other is the idea of uniformitarianism. Introduced by Scottish geologist James Hutton (1726–1797) in the late eighteenth century, it simply means that processes happening on the planet today are the same as those that happened millions of years ago. Like superposition, this idea seems simple and obvious, but it was revolutionary at the time. Many contemporary religious figures and some scientists adhered to the idea that most of the rocks on the planet were formed by a single giant flood or a series of cataclysmic planet-wide events – a theory called catastrophism.

As fossil finds became more common around the world, early paleontologists realized that

∧ Nicolas Steno was an early geological theoretician who recognized many of the principles on which modern geology is based.

> Mega exposures like the Grand Canyon in the USA allow direct examination of sedimentary processes. Clearly you can see that sedimentary rocks are lain one sequence above another and that the older rocks lie below the younger ones.

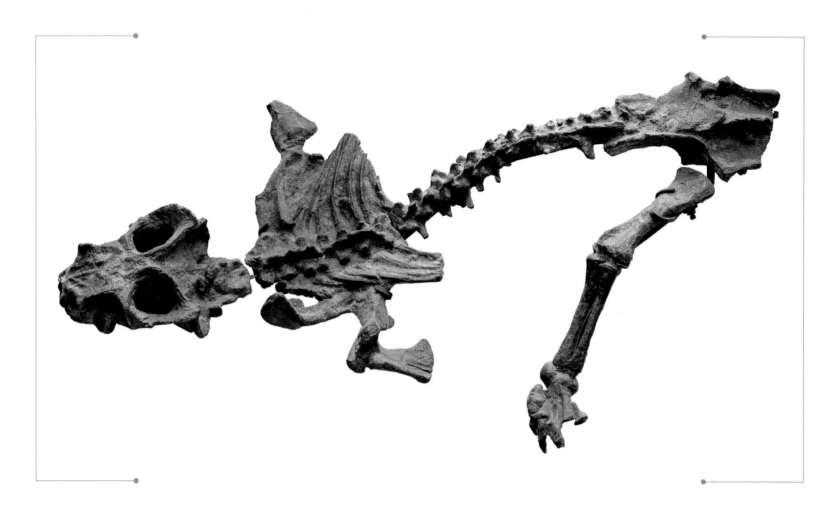

similar fossils were being found in different geographical areas. They correctly inferred that the rocks from which these fossils were excavated were the same age – or, as they termed it, "correlative". From this the sub-discipline of biostratigraphy was born.

By the late eighteenth and early nineteenth centuries, geologists settled on the idea of several basic divisions of geological time. Today, these are known as eons, eras, periods, epochs and stages, each representing a subdivision of the former. Through the work of European geologists, the eon during which advanced multicellular life evolved on the planet, the Phanerozoic, was divided into three subdivisions or eras. These are the Paleozoic, the Mesozoic, and the Cenozoic. Most of this book concerns the Mesozoic era, the time of the traditional non-avian dinosaurs. The Mesozoic is subdivided into three periods – Triassic, Jurassic and Cretaceous.

The Mesozoic era is punctuated by two of the largest extinction events in Earth history, the

Permian–Triassic event and the Cretaceous–Paleogene event. The Permian–Triassic event was the largest ever in Earth's history and it is estimated that up to 96 per cent of marine organisms and 70 per cent of land-living vertebrates disappeared. Although much is speculated, very little is actually known about the causes of this catastrophe. Many have suggested that it cleared the stage for dinosaur evolution by opening up terrestrial ecological niche space. Events surrounding the terminal Mesozoic event are much clearer; so clear, and of such consequence to dinosaurs, that it is examined in more detail in the final chapter.

The one thing that had escaped early geologists was the concept of time and the antiquity of Earth. Most estimates, like the famous one postulated by the Archbishop of Ussher in the seventeenth century and divined from the Bible, considered Earth to have been created in 4004 BC. Using a more empirical approach, in 1862 Lord Kelvin estimated the age of the Earth at between 20 and 400 million

∧ A specimen of *Lystosaurus*. Its remains have been found in Antarctica, South America and South Africa.

> Collecting in the Late Triassic rocks of the USA. Fossils from this interval look pretty similar worldwide – evidence for a single supercontinent of Pangea.

years. His calculations were based on the rate of cooling of the mass of the Earth from a molten orb to its present-day temperature. These calculations were further refined to give a date of 100 million years. Although a good try, Kelvin's approach was fraught with error.

Until the advent of radiometric dating, the age of the Earth was vastly underestimated. Even in the early twentieth century, the end of the Cretaceous period was thought to be only a few million years ago. Today the geological timescale is codified by an international association of geologists and it is firmly tied to real dates as opposed to stages and intervals. The age of the Earth is generally accepted as 4.5 billion years old.

Radiometric dating is relatively simple but filled with subtlety. Here, it will suffice to say

that it is based on the decay of radioactive elements found in the Earth. Often these are associated with fossils. Because the rate of decay of these elements into daughter elements (like the decay of Uranium into Lead) is known to be constant, the age of a rock unit can be determined by measuring the ratio of these elements. Several elements are used for dating rocks or fossils. Just a sampling beyond Uranium to Lead is one isotope of Argon decaying at a fixed rate into another isotope of Argon, the decay of isotopes of Carbon (Carbon 14 to 12), and many others. The mechanisms of how these clocks are set (whether volcanic, planet formation, or in the atmosphere) is a more complex matter, a discussion of which is not needed here.

The time periods of the Mesozoic –

Triassic, Jurassic and Cretaceous – are further subdivided into Early, Middle and Late, except for the Cretaceous, for which there is only Early and Late. The Triassic period ranges from roughly 252 million years ago to 201 million years ago. The period is named for three correlatable rock units occurring in north-western Europe, collectively called the "Trias". The Triassic was a time of great evolutionary experimentation. As noted above, following the Permian–Triassic event most terrestrial ecological niches were unfilled and vertebrates quickly claimed these with a number of new types. In the Early Triassic the dominant land animals were distant relatives of today's mammals and unusual crocodile-line creatures. It was not until the Late Triassic, around 240 million years ago, that dinosaurs first appeared. Their first occurrences were on a single supercontinent called Pangea; for the entirety of the Triassic all of today's continents were combined into this land mass. Although

beginning to fragment, Pangea remained a continuous entity until the Early Jurassic.

The environment during the Triassic period was largely hot and dry across Pangea. However, there is some evidence, especially late in the Triassic, that monsoon systems had developed, leading to more humid environments in some sub-regions. Forests would have looked very different to today, although some plants species that still survive were present. These include conifers, ferns, relatives of the gingko, and horsetails. The ocean was populated by a variety of fishes as well as marine reptiles such as ichthyosaurs and plesiosaurs – all of which would disappear by the end of the Cretaceous period.

The Jurassic period, which began about 200 million years ago and ended around 152 million years ago, is the time interval where dinosaurs proliferated and diversified into all of their major subgroups. Named after the Alpine Jura mountain range, Jurassic dinosaur

> Maps of the Mesozoic era show the dramatic shifts in continental arrangement that occurred during this time interval.

∨ Many fossil plants (like this fern) and animals are distributed across several continents. This is evidence that those continents were once annealed into a single larger continent.

localities are strewn across the globe. Other reptile groups, such as lizards, turtles, and crocodiles, also diversified and developed a characteristic modern look. Modern sorts of mammals evolved, and the first flying dinosaurs, the birds, also appeared. Forests became much more familiar to our eyes. Even though angiosperms had yet to appear, many modern sorts of plants were present.

Pangea began to separate, first into northern and southern continents called Laurasia and Gondwana respectively. This fragmentation, along with the escalation caused by the co-evolution of plants and animals, led to extreme diversification, especially among herbivores. By the Middle Jurassic some amazing dinosaurs were present, including some of the largest animals ever to occupy the terrestrial realm – the sauropods. By the end of the Jurassic, the animal fauna had a decidedly modern feel to it.

The Cretaceous period, from about 141 million years ago to approximately 65 million years ago, is when an explosive plant radiation occurred and angiosperms became the dominant plant group on the planet. Almost all of the plant groups which are consumed today by herbivores (including ourselves) are angiosperms. As far as we can garner from the record, it is during this interval that dinosaurs reached their greatest diversity. This is not surprising because it is the time period where the continents were at their most fragmented. By the end of the Cretaceous period, the continental areas were roughly in the geographical positions that they are today, but sea levels were much higher and both North America and Eurasia were bisected by huge shallow intercontinental seaways. Southern Europe was similarly inundated and existed as a tropical island archipelago. As with such geographies today (the Galápagos islands, Indochina), these insular areas were incubators for speciation, and it is no surprise that the dinosaur faunas of eastern North America are quite different from the faunas of western North America. This immense diversity and disparity among dinosaurs in the Late Cretaceous came to a crashing end about 65 million years ago. All that is left are the birds around us today – represented by some counts at about 18,000 species.

THE BREAKUP OF PANGEA

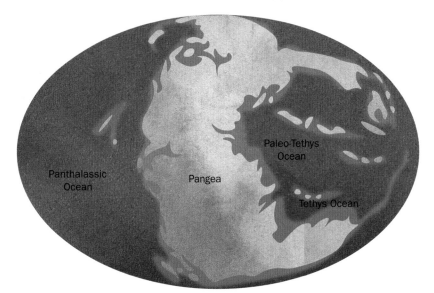

TRIASSIC

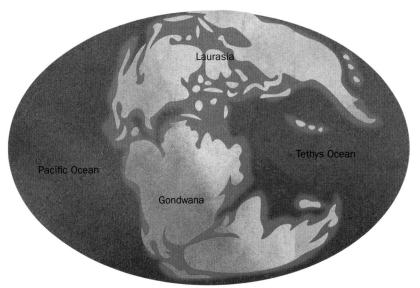

LATE JURASSIC

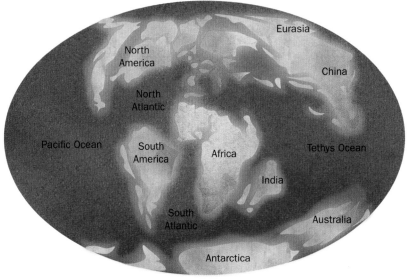

CRETACEOUS

DISCOVERING DINOSAURS

HOW ARE DINOSAURS FOUND? THIS GENERALLY HAPPENS IN ONE OF TWO WAYS. THE FIRST IS OBVIOUS: PEOPLE ARE LOOKING FOR THEM. USUALLY THESE ARE PROFESSIONALS (BOTH ACADEMIC AND COMMERCIAL) ON PLANNED EXPEDITIONS. PERMISSION IS ALWAYS REQUIRED, EITHER FROM GOVERNMENTS, BOTH LOCAL AND NATIONAL, OR, IN THE CASE OF PRIVATE LAND, FROM THE LANDOWNERS. THE OTHER IS THROUGH HAPPENCHANCE; MANY IMPORTANT SPECIMENS HAVE BEEN DISCOVERED BY LUCKY AMATEURS OUT FOR A HIKE OR CYCLE RIDE, OR THROUGH CONSTRUCTION WORK.

Professional paleontological exploration techniques have not changed a great deal since the early expeditionary days in the field. Yes, we now have satellite imagery, GPS, and, in some cases, drones and helicopters. While new technology has made it easier to identify places to search, there is no substitute for spending a lot of time on the ground. Strong legs, and a willingness to hike long distances are the primary requirements for this type of work. Not everyone enjoys this and not all dinosaur paleontologists actually undertake this sort of fieldwork.

Usually potential sites are identified due to three reasons. The most obvious is that there is a particular question that is in need of answer. In the early days, there was a need and consequently a rush to fill museum galleries with fossils, but this is certainly not the case today, and fieldwork is more question-driven. In my case, it has to do with the transition from traditional dinosaurs to birds. So, areas and time intervals that preserve these kinds of fossils are where we look. That flows into the second way to identify a potential site; someone has previously found something in that area. A fossil locality is never as good as it is on the first day of discovery, but many localities that are over 100 years old continue to offer up quality specimens. Finally, the most difficult but sometimes the most rewarding expeditions go to where no one has looked before. Reasons for non-exploration run the gamut from extreme remoteness to political reasons (permits, social issues, and even war and lawless societies).

Once an area is identified, plans are made and usually a small scout party travels to the region. In most cases these trips go lean and light as it is not a good idea to spend large sums of money or a lot of time in an area where there are no fossils of interest. If the initial survey goes well, plans are made, money is raised, personnel are identified and recruited, and infrastructure is put in place and arranged.

Once at the locality, specimens that are worthy of excavation are identified. Resources and time do not allow everything to be excavated. The excavation strategy is determined by the remoteness of the locality, the size of the specimen, how much rock

> A Museum field crew excavating at Ukhaa Tolgod in Mongolia's Gobi Desert.

needs to be removed and the hardness of the sediments. Sometimes specimens are found on cliff faces, requiring ropes and rigging. Dinosaur fossils are even found on seashores where there is only a narrow window between tides when they can be worked on.

Exposed bones are first consolidated with special glues that are of the type used by art conservators and which can be removed. More and more of the sediment is exposed until there is a deep ditch around the specimen. Layers of soft paper are then applied to the block where the bones are exposed. Strips of burlap or sacking soaked in plaster of Paris are used to consolidate the entire block in a neat package known as a jacket, much in the same way as a broken bone would be set in a cast. Chisels are driven under the specimen and the jacket is flipped over with the specimen hopefully remaining in it. Then the process of paper, burlap and plaster is repeated again, completely enclosing the specimen and the rocks entombing it.

After the specimen arrives at the lab, skilled technicians called preparators open the jacket. Using a variety of tools, ranging from miniature sandblasters and jackhammers, to tiny needles and chisels, the rock is removed from the matrix. This is somewhat facilitated these days since the blocks can be X-rayed and, in some cases, CT scanned as an aid to preparation. A support of Styrofoam is created and the dinosaur fossil is ready for study or, in some cases, after mounting, for exhibit.

In the last few decades dinosaur fossils have become a commodity. While fossil sales were increasing throughout the 1990s, it was the unprecedented sale of a *Tyrannosaurus rex* specimen called "Sue" in 1997 that really changed everything. This specimen has an unusual, unfortunate and convoluted history.

It is the most complete *Tyrannosaurus rex* specimen yet collected. It was found on private land in the USA. In the US, unlike most countries, fossil specimens collected on private land are the property of the owners of that land. In other countries, fossils (and archeological artefacts) are the property of the

> Museum paleontologist Walter Granger at Bone Cabin Quarry in Wyoming, one of the early Museum excavations.

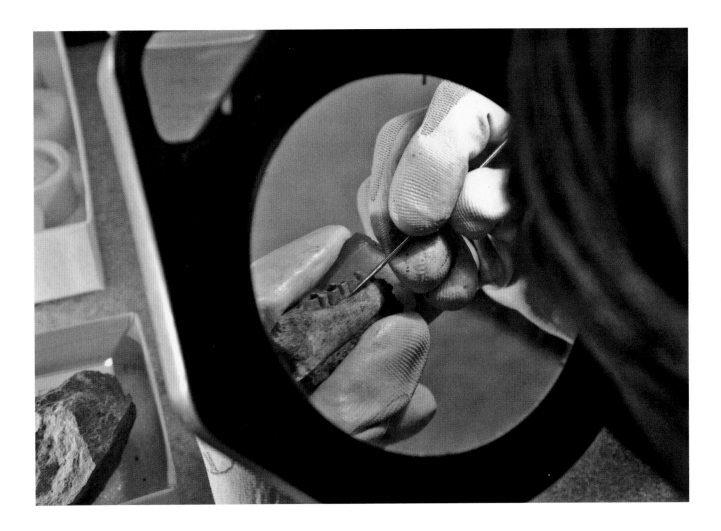

government. The fossil was named "Sue" after the woman who found it. She was a member of the commercial paleontological company that excavated the specimen. Because of a contractual dispute between the land owner and the fossil excavators, extremely contentious court proceedings followed, after which the fossil was released to the land owner.

Feelers were put out for a purchaser and no one came forward. Eventually it was decided that the specimen should go to auction. In October 1997, Sotheby's auction house put the Sue specimen on the block in New York City. The results stunned everyone. The upside estimate was just north of $1 million. Bidding ended with hammer price plus commission at $8.36 million in a sale that was publicized worldwide. The specimen was purchased by the Field Museum in Chicago, USA. This purchase sent shudders through the dinosaur community and changed the world of dinosaur paleontology forever. No longer were paleontological specimens simply the subject of eccentric paleontologists,

museum-goers, children and nerds. They were now a commodity.

This is good and bad. Raising the profile of these objects can be a good thing in that institutions have recognized the value (not just in a scientific sense) these specimens hold. But it has created issues. Land owners in fossiliferous regions are no longer as willing to allow institutional paleontologists on their land as they commonly did in the past. Instead they lease their property to commercial operations. The monetization of fossils has also led to extreme levels of poaching around the world.

The most famous case was the attempted sale of a close *T. rex* relative, an animal called *Tarbosaurus baatar*, at Heritage auction house in New York City in 2012. This specimen was featured on the cover of an auction catalogue and was recognized as a specimen that had been smuggled out of Mongolia against Mongolian law. Although there were repeated requests by the Mongolian government to prohibit the sale, Heritage proceeded on schedule even though the Mongolians had

retained a US attorney to try and halt the sale. The specimen sold for a little under a million dollars, but it was never collected by the buyer. Instead the legal machinery worked and the specimen was seized by the federal government and returned to Mongolia's capital, Ulaanbaatar, where it is now on view. Sadly, poached important specimens from Mongolia, China and the USA continue to be traded on the dark commercial market.

∧ After collection, and before study, fossils must be removed from their matrix. This work, called preparation, is painstaking and tedious. It requires the eyes and hands of skilled technicians.

< The Sue specimen at the Field Museum in Chicago. It is the most complete *Tyrannosaurus rex* specimen yet discovered.

DINOSAUR BIOLOGY

WE CANNOT DIRECTLY OBSERVE AND STUDY TRADITIONAL DINOSAURS.
WE CAN'T TAKE THEIR TEMPERATURES, MEASURE THEIR HEART RATE, TAKE
THEIR BLOOD PRESSURE, WATCH THEM COPULATE AND LAY EGGS, MEASURE
THEIR SPEED OR LUNG CAPACITY, SEE HOW MUCH THEY EAT, OR HOW THEY
BEHAVE, HOW THEY ATTRACT A MATE, OR HOW THEY DEFEND TERRITORY.

Nevertheless, we have been able to determine a great deal about them using a variety of means. The most powerful of these is an empirical application of the comparative method. Through this we can study animals alive today to make powerful informed inferences about organisms that have been dead for millions of years.

This works by using living animals and the relationships among them to predict features that do not necessarily preserve as fossils. The closest relatives to traditional dinosaurs are living birds (see "Birds and Dinosaurs", p.224). Birds are a type of dinosaur and are reptiles, just like humans are both primates (great apes and monkeys) and mammals.

In the chapter "What Are Dinosaurs" the family tree of archosaurs was discussed, and the primary division between the crocodile-line and bird-line lineages and their respective diversity defined. The closest living relative to birds are crocodilians and this relationship can be used to understand much about the way dinosaurs lived.

A simple parallel example: the closest living relative to humans is the chimpanzee. We share lots of characteristics with chimps and interpret these as being present in our common ancestor and we retain these features by descent. We do not postulate that they evolved independently.

Hominids such as *Australopithecus* are more closely related to humans than chimps, but we don't know a lot about them since all we have are a few meagre fossil skeletons. No hair is preserved, but every interpretation shows them covered with varying amounts of hair. Why? Because *Australopithecus*, although more closely related to humans than to chimps, evolved from the mutual common chimp-human ancestor. In lieu of other evidence, we predict that all animals that shared this common ancestor had hair.

We can look at crocodilians and birds in the same way and find characteristics that they share and predict their occurrence in non-avian dinosaurs. A few good examples are eggs laid in a nest (no crocodilians or birds are live-bearing), an advanced circulatory system (both of the extant groups have a four-chambered heart allowing a more active lifestyle than other reptiles), and efficient lungs where air passes in one direction through the gas-exchanging membranes. All of this, and more, can be determined to have been present in non-avian dinosaurs without looking at a single dinosaur fossil.

Other data sets directly use fossils to inform us about dinosaur biology. Analysis of this information has shown us much about how these animals lived. Because dinosaurs are such a morphologically diverse group (ranging from tiny flyers to the largest terrestrial animals), their biologies would be expected to match this diversity. Entire books are devoted to the topic. But we will take a brief look at a little of this here and talk more about it as it applies to specific animals in subsequent chapters.

∧ The lungs of archosaurs are a revelation. Big air sacs store air and, unlike humans, the inhale (fresh air) does not mix with the exhale (air diminished in oxygen). This is portrayed in a Museum display where the air sacs are highlighted as coloured areas.

Behaviour

Crocodilians and most birds are social animals. Crocodilians have been known to hunt cooperatively and they guard their nests. It is predicted that such behaviours would be present in non-avian dinosaurs. There is plenty of evidence for social behaviour in these animals. Fossil trackways show herds of sauropods walking in concert, and mass death assemblages, where entire flocks of dinosaurs were killed by a single event, are preserved. Much is predicted about the hunting prowess of specific theropods based on their teeth and engineering analyses of their skull and skeleton. Feeding in herbivorous dinosaurs is also well understood. These studies have also focused largely on the teeth and are especially developed in ornithischian groups such as hadrosaurs (duck-billed dinosaurs) and ceratopsians (horned dinosaurs).

Because non-avian dinosaurs are such a diverse group, they undoubtedly displayed a huge variety of behaviours and feeding preferences and strategies, probably commensurate with what we see in mammals today. As with other aspects of dinosaur biology there is voluminous literature on this subject, and still we have barely scratched the surface.

∧ Living crocodilians (as shown by this alligator) build and guard their nests until the young hatch. In some cases, the mother digs the newly hatched young out from the vegetation-covered nest.

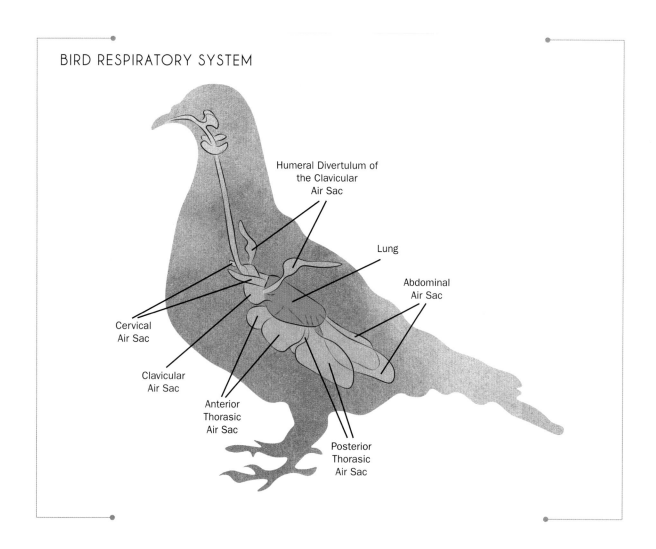

BIRD RESPIRATORY SYSTEM

Humeral Divertulum of
the Clavicular
Air Sac

Lung

Abdominal
Air Sac

Cervical
Air Sac

Clavicular
Air Sac

Anterior
Thorasic
Air Sac

Posterior
Thorasic
Air Sac

Physiology

All of the evidence is pointing to the idea that dinosaurs were intelligent and active. Much evidence indicates that they were warm blooded, although not all of them to the level of homeothermy present in living birds. The evidence comes from the fact that their respiratory system was so advanced. In our lungs we breathe in and out. If our lungs were 100 per cent efficient, we would extract 50 per cent of the oxygen in the air and dump 50 per cent of the carbon dioxide in our bodies back into the atmosphere. But our lungs are only about 5 per cent efficient in removing oxygen from the air, while bird and crocodile lungs are much better, probably exceeding 25 per cent of the available oxygen.

In birds and crocodiles, the air is inhaled into a series of air sacs, not directly to the lung. From these air sacs it then goes to the lung where gases are exchanged, before being discharged to another set of air sacs and then

out. What is unique about this system is that the gas-exchange membrane is a counter current type. Simply, this means that when the blood from the body, laden with carbon dioxide and low in oxygen, enters the lung it first encounters air from the outside that is rich in oxygen and relatively low in carbon dioxide. This is a perfect condition for a very efficient diffusion gradient as oxygen flows into the blood and carbon dioxide into the air. By the time the blood reaches the other end of the lung it is rich in oxygen and low in carbon dioxide, while the exhaled gases are rich in carbon dioxide and low in oxygen.

Large air sacs also have implications for the mass of these animals because they are closely associated with, and permeate, the bones of birds. Air sacs leave tell-tale signatures on bones in the form of large cavities and hollows. In extinct dinosaurs this is poorly understood, however in living dinosaurs, these air sacs both work with the lungs and lighten

∧ The intricate lung of a bird showing the air sacs. All these sacs work together to form the most efficient respiratory system of any vertebrate.

> A juvenile *Hypacrosaurus*. Over the last two decades we have learned much about dinosaur growth and reproduction.

the skeleton. Marks indicative of air sacs are known on lots of dinosaurs but are primarily apparent in the saurischians. And this is one of the main reasons that the estimation of dinosaur mass (especially saurischians) is so difficult. A good example is to compare an adult chicken and a puppy. Both have relatively the same volume, yet the puppy will generally weigh twice as much or more than the chicken. The puppy is flesh and bone; the chicken is flesh and bone punctuated by large air sacs decreasing its density.

One of the more controversial areas of dinosaur physiology is the question of warm bloodedness. First proposed in the early 1970s, this remains a contentious issue. Among extant animals, only birds (living dinosaurs) and mammals have the ability to raise their temperature above ambient. Many characteristics of living birds, such as air sacs, respiration, and a four-chambered heart, facilitate this metabolism. Also, feathers almost

certainly work as thermal blankets keeping animals warm in the same way our hair does.

But there is other evidence as well. In one notable study an entire dinosaur quarry was excavated. Every single bone was identified and classified as to whether it was a warm-blooded animal (a mammal) or a cold-blooded one (crocodilian, lizard, turtle, etc.). Lots of different sized animals were collected. There are techniques that allow scientists to approximate the temperature of a dead or fossil animal when it was alive using ratios of chemicals preserved in their bones. What was found was that turtles and crocodilians showed a lot of variability in temperature when they were small, only becoming more stable as they became larger. That is because as an animal gets bigger it holds onto heat better. At home you see this when it takes longer for a 6kg (13¼lb) turkey to freeze in a freezer than a 500g (17½oz) Cornish game hen. What they found for dinosaurs (and these were hadrosaurs

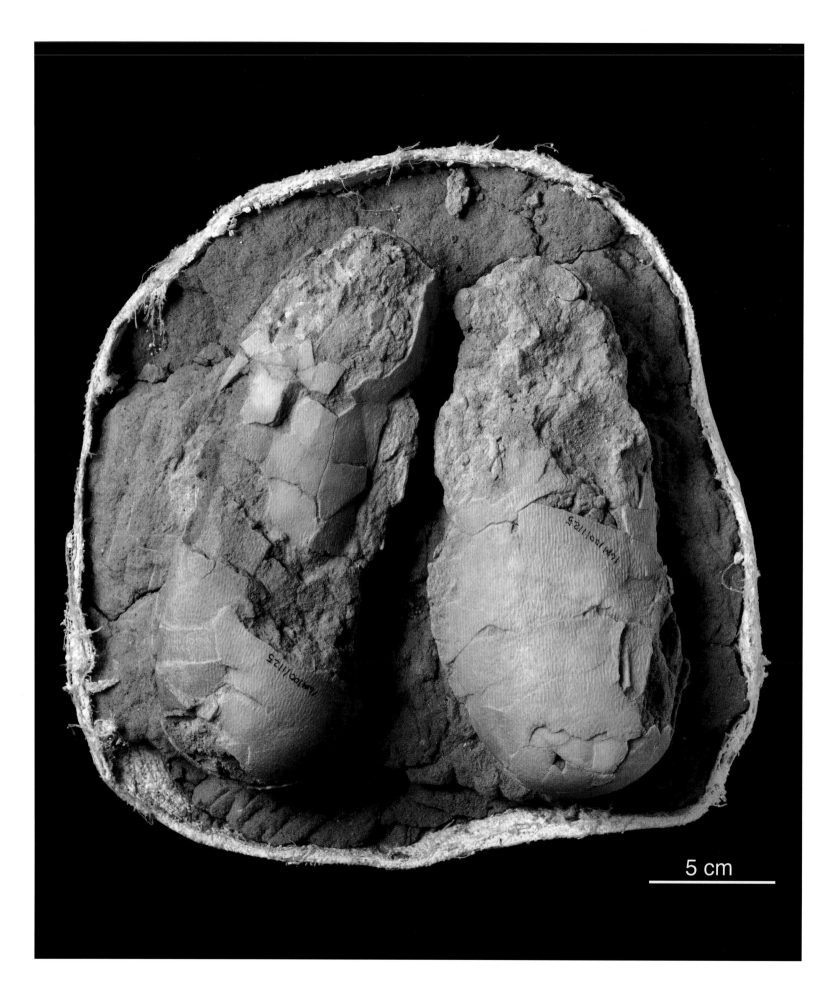

5 cm

< The paired arrangement of these oviraptorid dinosaur eggs from the Gobi Desert suggests that they were laid two at a time.

∧ Looking at very thin slices of dinosaur bones can tell us how old they were. In this specimen even the cavities that enclosed small bone cells are preserved.

– remotely related to modern birds) was that they were able to keep a stable temperature no matter how much they weighed. This is called thermal inertia, and we see it in the heating process – large ice blocks melt more slowly than small ones – and the cooling process. Often in the morning large rocks are still warm from the heat of the previous day, while small ones made of the same material are cool. This is powerful evidence that these animals had the ability to create metabolic heat. Unless they generated their own heat, the small dinosaurs would have had highly variable temperatures like the cold-blooded lizards, crocodilians and turtles. They didn't. Their chemical temperature signature was just like the larger adults, suggestive that they were warm blooded to some extent.

There are other lines of evidence that will be discussed in later chapters. Most of this is anecdotal (like behaviour, how fast animals grew, and where they lived). Nevertheless, it all points in the direction that non-avian dinosaurs were capable of producing metabolic heat and could maintain temperatures well above ambient.

Reproduction

We know more about dinosaur reproduction. Dinosaurs laid their eggs in nests. These nests are found both as solitary groups of eggs and as communal associations. The eggs of dinosaurs that are more distantly related to birds – hadrosaurs and sauropods, for instance – are nearly round and their pores, which allow gas exchange with the atmosphere, are uniformly distributed across the egg's surface. Eggs of animals more closely related to birds are elongate orbs or asymmetrical and have pores concentrated at one pole.

Nests of distant bird relatives are poorly organized, random accumulations. Occasionally it can be determined that the eggs show some degree of pairing. This is because female dinosaurs laid two eggs at one time; one from each oviduct. One of these oviducts was lost near the origin of modern birds, as only one functional oviduct remains in living species. Somewhere within Theropoda, eggs became organized into proper nests. These are particularly apparent in dinosaurs of the group Oviraptoridae.

Growth

Non-avian dinosaurs grew fast – not as fast as birds but faster than their proximate relatives, the crocodilians. As you get closer in the family tree to birds, the faster they grew. A surprising fact about non-avian dinosaur communities is that very few animals reached full adulthood. We can determine this on the basis of their bone histology, meaning that if we look at cross sections of bones we can determine how old the animals were in the best cases.

One of the best studies considers growth in *Tyrannosaurus rex*. *T. rex* grew very fast. There are two ways that an animal can become larger than its ancestor. One is to grow faster, the other is to grow longer. This is something that we can directly measure because growth is recorded in the bones of these animals as annual lines. Called "lines of arrested growth or lags", these rings can be seen through a cross-section of the bone, analogous to tree rings. If the animal is growing quickly, lags are widely spaced. When an animal reaches full size and stops growing, lags are very close together and sometimes hard to distinguish. *Tyrannosaurus* has many smaller sized close relatives, such as *Albertosaurus*. If we compare lags between these animals and other tyrannosaurs, we see that all of them reached full size at about 18 years. The difference is that between 13 and 19 years *Tyrannosaurus* grew explosively up to 2kg (4½lb) per day.

Biomechanics

Sophisticated computer simulation has allowed the speed of *T. rex* to be accurately calculated. Virtual skeletons can be modelled and muscles attached to the bones. Vertebrate bones have rough patches that indicate where muscles attach. Since *T. rex* basically has the same anatomy as a chicken, except for the tail, these rough patches correspond with known chicken muscles. The angles of movement between joints are then put into the model along with values for the strength of the muscles. Muscle strength is calculated on the basis of the cross-section of the muscle at its widest point. A variety of values are used from small to large, but all within reason. This is called

∧ *T. rex* had formidable teeth. These animals were able to generate the greatest bite force of any known animal — so large, that the force could have exploded solid bones.

< Scientists who study biomechanics can reduce dinosaur skeletons to engineering forms. They use these to determine animal speed and posture.

a sensitivity analysis and results in a range of running speeds dependent on the muscle sizes used. When the model is run, the speed can be calculated. For *T. rex* the top speed was about 12mph (19kph). Because of the weight of the animal, top speeds could not be achieved for long durations, and even at full speed an active human could out run it.

Such modelling also allows exploration of the feeding mechanics of *T. rex*. It has been calculated that *T. rex* had the greatest bite force known of any creature – over 3,622kg (7,895lb), the weight of a small truck and exceeding the weight of two Mini Coopers. The force was so great that if it penetrated bone, the bone would literally explode. This is unique, and points to an uncommon feeding strategy where the prey was dispatched with trauma. This is evidenced to some extent by the presence of many splinters of bone in coprolites (fossilized dinosaur faeces), indicating that the bones were fragmented at ingestion.

Studies in this detail have yet to be applied to many other dinosaur species, but they are beginning to give us a better picture of these incredible animals.

DINOSAUR CLASSIFICATION

THE CHAPTER "WHAT ARE DINOSAURS?" COVERED THE GENERAL ORTHODOXY SURROUNDING THE FAMILY TREE OF DINOSAURS. ALTHOUGH THERE HAVE BEEN RECENT CHALLENGES, THESE HAVE NOT BEEN SUFFICIENT TO OVERTURN THE PREVAILING CANON ESTABLISHED IN THE NINETEENTH CENTURY THAT DINOSAURS ARE OF TWO TYPES: SAURISCHIANS AND ORNITHISCHIANS.

At least 250 million years ago, Archosauria split into two: one group was bird-like, including dinosaurs, pterosaurs and birds; and the other crocodile-like, with alligators, crocodiles, and many now-extinct relatives.

In the modern world many regard crocodiles as loathsome, primarily aquatic carnivores. Their early history shows a much greater diversity. Many of these animals were fully terrestrial quadrupeds which were top-level carnivores. *Prestosuchus*, for instance, lived in what is now Brazil, was the size of a tiger and was definitely a top-level carnivore. Others, such as the aquatic phytosaurs, filled similar roles to crocodilians today.

Other crocodile-line archosaurs include animals such as *Effigia*, which lived in the Late Triassic in what is now New Mexico in the USA. It was toothless and bipedal. It probably filled an ecological role similar to that of ostrich dinosaurs (ornithomimids).

Some animals in this lineage looked like pangolins or armadillos. Named the aetosaurs, they sported bony armour and a few spikes on their bodies. Never numerous, and not very large (only about 1m (3¼ft) in length), they are enigmatic. They had tiny teeth and little is known about diet or behaviour.

As we move closer to modern crocodile

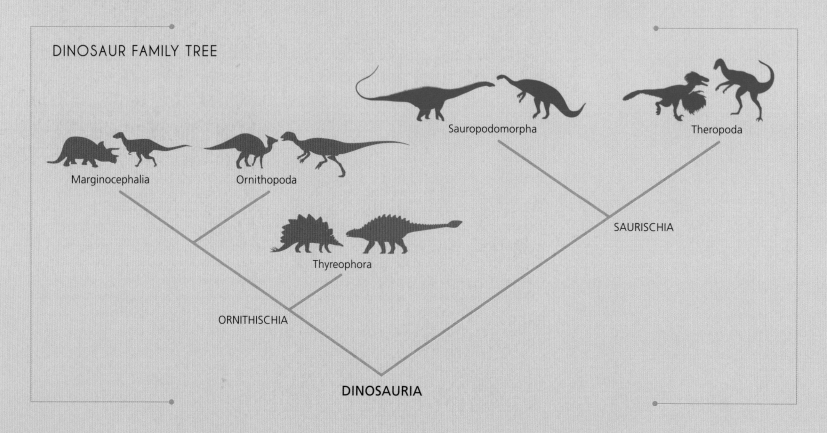

DINOSAUR FAMILY TREE

Marginocephalia

Ornithopoda

Thyreophora

ORNITHISCHIA

Sauropodomorpha

Theropoda

SAURISCHIA

DINOSAURIA

diversity there are some amazing animals. One group, the metriorhynchids, were marine animals. These creatures were large, formidable predators that filled an ecological niche which in today's world is filled by the great white shark.

Even within groups of animals that are quite closely related to the diversification of the 23 living species of crocodilians there is a fair amount of disparity. There are tiny wholly terrestrial spiky forms (gobiosuchids) which had limbs that are the diameter of a piece of spaghetti. There are giant toothless forms with skulls nearly 2m (6½ft) long. These were aquatic and had skulls so long and flat that they resembled surfboards. In what is now South America and the US state of Texas, large forms existed with skulls over 1m (3¼ft) in length. In Africa, extremely large forms (11m/36ft) existed. It has been suggested that the African and North American species, *Sarcosuchus* and *Deinosuchus*, fed on dinosaurs. These animals were giants with estimated lengths approaching 18m (59ft). Other bizarre crocodilians (the Pristichampsidae) had hoofs and blade-like teeth and were top carnivores for a relatively short interval after the traditional dinosaurs became extinct. There were even crocodilian species that developed very mammalian, multi-cusped teeth and others that had giant bulbous teeth specialized to feed on clams or turtles.

Today's 23 living species of crocodilians represent a vestige of this diversity. They all look much the same in having a very conservative body plan. The largest is the saltwater crocodile (*Crocodylus porosus*), which can reach a length of over 6.2m (20⅓ft). All of them (except species of *Alligator*) live in a circumtropical belt around the planet.

The bird line, also descended from the first archosaur, is just as complex. Its early history is not well understood. There are various poorly preserved fossils of animals that fit the early history of this lineage, but the most relevant is the immediate most diverse group, which are close dinosaur relatives – the pterosaurs. Often improperly labelled "pterodactyls" and also "flying dinosaurs", this group was incredibly successful. They were the first vertebrates to achieve true powered flight. First appearing in the Late Triassic their fossils have been found worldwide. They range in size from tiny animals smaller than a house sparrow, to giants with a wingspan that approaches 20m (65½ft) – the size of a small aeroplane. However, their skeletons are so specialized that they do not tell us much about the ancestral dinosaurs. Yet it would be amazing to see them glide, hunt, flap and soar through the skies.

To recap, Dinosauria (despite recent challenges) is divided into two groups, Saurischia and Ornithischia. Here we will lay out the ground plan for the genealogy of the animals that we will explore in more detail in later chapters. (See also, the family tree on page 53.)

For the purpose of this book, the saurischians are composed of sauropods (gigantic dinosaurs)

and theropods. Primitive sauropods are called prosauropods. This may not be a monophyletic group, as some may be more closely related to more advanced members than others. Some like *Plateosaurus* are more primitive, others like *Anchisaurus* are probably more closely related to more advanced animals such as *Diplodocus*. More primitive sauropods were bipedal, with relatively short necks and tails. More advanced sauropods had long necks and long tails. Members of this group were the largest animals ever to walk on land. All were thought to be herbivorous.

Theropods were for the most part bipedal, yet a few, such as *Spinosaurus*, may have been secondarily quadrupedal. Most members are poorly understood but include animals such as *Herrerasaurus* and *Tawa* – small bipedal cursors (rapid runners). The next level of hierarchy is the Neotheropoda, the most primitive of these are the coleophysids. For the species covered in this book, *Coelophysis* is a good example (see p.56). As we move to more advanced theropods the next group is called Averostra, and the first group to diverge off of this line is the Ceratosauria, which includes

such animals as the eponymous *Ceratosaurus* and the primarily southern hemisphere abelisaurs. All of these were large bipedal carnivores and many sported elaborate head crests and horns.

As we move up the family tree, the animals become both more familiar and better known. The large and diverse group called the Tetanurae includes most of the best-known theropods as well as birds. The most basal dinosaurs in this group are the megalosaurs; *Megalosaurus* was one of the first dinosaurs to be recognized. The megalosaurs include some of the most peculiar identified dinosaurs. Also known as spinosaurs, they measured up to 14m (46ft) long, sported huge sails on their backs, and had long muzzles with a mouthful of pointy teeth. They are thought to have occupied littoral habitats and fed on fish.

When we reach Avetheropoda the animals become even more advanced. The first group to branch off here are the Allosauroidea. In general, this group is widely distributed but not particularly well known, except for *Allosaurus* (see p.64). Beyond the Allosauroidea is the crowning

group of Theropoda, the Coelurosauria.

Coelurosaurs include a huge range, from *Tyrannosaurus rex* (see p.72) to hummingbirds, incorporating all feathered, bird-like animals and true birds. They had big brains, advanced behaviours, and evolved almost all the traits that most perceive as bird characters.

The other side of the tree, the Ornithischia, is similarly complex. At the base are small, primitive, bipedal dinosaurs, such as *Heterodontosaurus* and *Lesothosaurus*. Like all ornithischians they were herbivores. It was not until the Early Jurassic that more derived forms begin to radiate and diversify. Collectively these animals are called Genasauria and the first branch to split off were the thyreophorans.

Among dinosaur groups Thyreophora is one of the most perplexing. The members of this group are known for their external accoutrements, namely body armour, spikes, plates and tail clubs. Some grew very big, and a paleontological conundrum exists over how to reconcile such large animals with their preposterously small mouths and feeble teeth. These unusual animals had a global distribution and are typically divided into two groups. One of these, the stegosaurs, are often called plated dinosaurs. This is something of a misnomer because some species did not have plates, but instead sported large spikes. The ankylosaurs comprise the other group. In general these were large, lumbering tank-like animals; their most primitive members were small forms. Thyreophorans have a good but puzzling fossil record in that, while we know a great deal about their morphology, we know relatively little about what their plates were used for, or their diets.

Moving up the tree to more advanced forms we have a group called the Cerapoda. Within Cerapoda is the great flowering of ornithischian diversity. As with herbivorous animals today, these creatures evolved in concert with evolving geographies and plant evolution. They are very numerous. Just as today in any community there are more kinds of herbivorous animals than carnivores, so it was in the Mesozoic era. Cerapods include two ubiquitous and familiar kinds of dinosaurs.

One group is Ornithopoda. They were named ornithopods in the nineteenth century because their feet have three primary toes that all point forwards, superficially resembling the condition in modern birds, even though they are not closely related. Primitive ornithopods are often referred to as "hypsilophodontids" and include such animals as *Thescelosaurus* and *Hypsilophodon* (see p.206–209). While very common and interesting because of unusual features such as weird teeth and large eyes, animals like *Hypsilophodon* are considered more closely related to the more advanced ornithopod group, the hadrosaurs, than *Thescelosaurus*.

Hadrosaurs are one of the most fascinating groups of dinosaurs and are unmatched in their diversity and uniqueness from other dinosaurs. They are known from dozens of species, almost all with a large flattened muzzle (leading to their nickname "duck-billed dinosaurs") and many sported elaborate crests on their skulls. Although the use of these crests in life in specific hadrosaurs is hotly debated, it is almost uniformly agreed that they played some display function, similar to horned animals like antelope do today.

< A technician dusts off a skeletal mount of spinosaur *Suchomimus*, a truly bizarre animal. The diversity of dinosaur body forms never ceases to amaze even professionals with every new discovery.

∧ A small bat-sized pterosaur from the Early Cretaceous of China. This one is covered with fuzz that may be related to dinosaur type 1 feathers. Pterosaurs are close dinosaur relatives but are not dinosaurs.

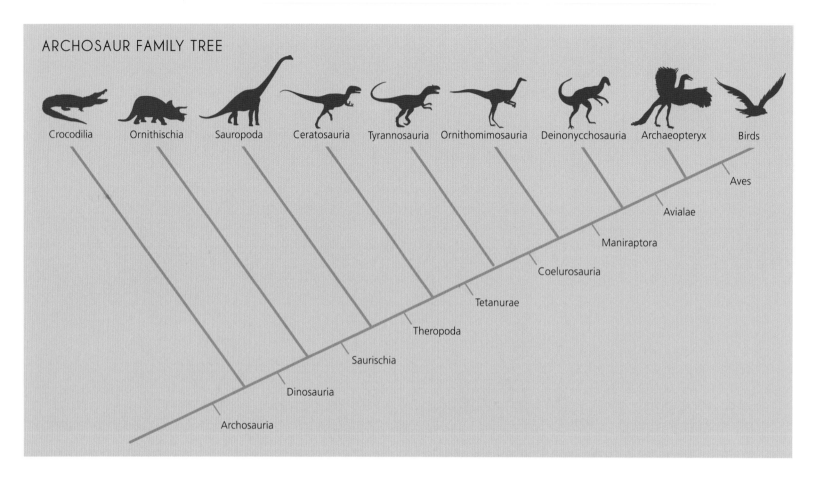

ARCHOSAUR FAMILY TREE

Crocodilia · Ornithischia · Sauropoda · Ceratosauria · Tyrannosauria · Ornithomimosauria · Deinonycchosauria · Archaeopteryx · Birds

Aves

Avialae

Maniraptora

Coelurosauria

Tetanurae

Theropoda

Saurischia

Dinosauria

Archosauria

The final group of cerapods are the marginocephalians. This group is formed of strange, dome-headed dinosaurs – the pachycephalosaurs; and the horned dinosaurs – the ceratopsians. Collectively these are known as Marginocephalia, referring to the bony outgrowths on the perimeter of the back of the skull. Like hadrosaurs, this group was successful, fairly cosmopolitan and very diverse, and they also exhibited many strange structures on their skulls. Although they look fearsome, the spikes and frills would have proven little defence against a megapredator. Instead they were most likely objects of display; whether for recognizing their own species, frightening predators or attracting mates, these head ornaments are unmatched in the dinosaur world.

∧ A very simplified archosaur family tree.

> Some pterosaurs became giants. The neck vertebra on the right comes from an animal that may have had a wingspan in excess of 17m (55¾ft).

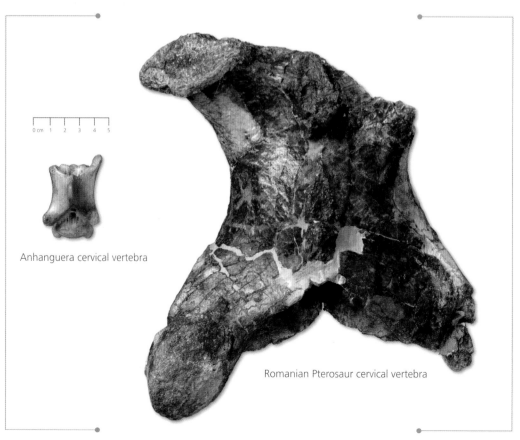

0 cm 1 2 3 4 5

Anhanguera cervical vertebra

Romanian Pterosaur cervical vertebra

PHYLOGENETIC/FAMILY TREE

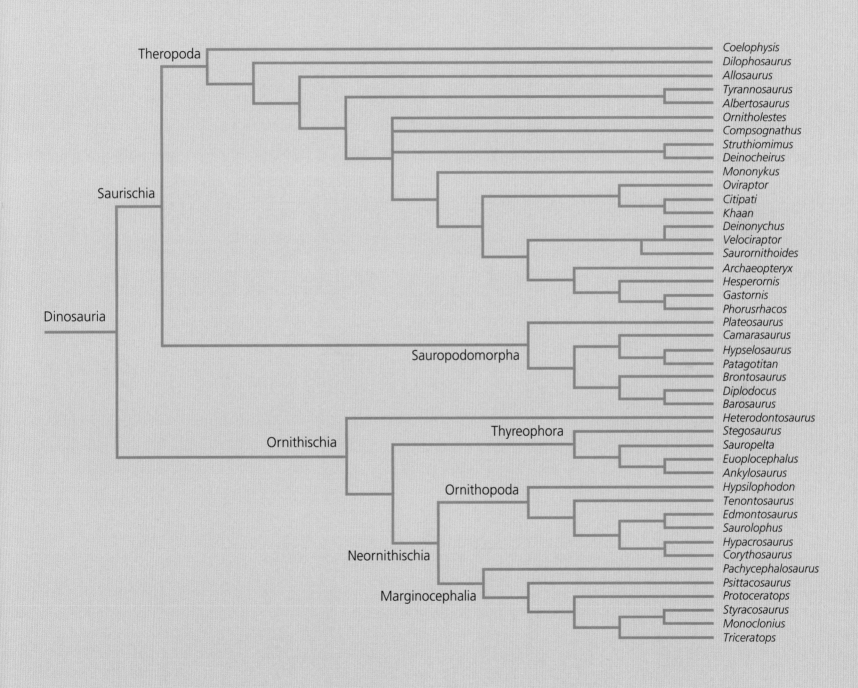

SAURISCHIA

THEROPODA
56

SAUROPODOMORPHA
136

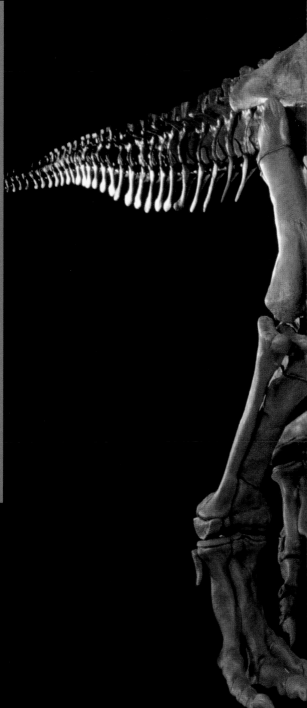

COELOPHYSIS BAURI

LATE TRIASSIC
CHINLE FORMATION
SOUTHWESTERN UNITED STATES

COELOPHYSIS WAS ONE OF THE FIRST PRIMITIVE THEROPOD DINOSAURS TO BE RECOGNIZED. THE NAME LITERALLY TRANSLATES AS "HOLLOW BONE". THIS IS BECAUSE THE FIRST SPECIMENS TO BE DISCOVERED SHOWED A FEATURE CHARACTERISTIC OF THEROPOD DINOSAURS, INCLUDING LIVING BIRDS, OF BONES CONTAINING LARGE AIR SPACES.

The first specimens found were highly fragmentary. These were collected by David Baldwin, a talented member of E.D. Cope's teams in the 1870s in Northern New Mexico, in what was at the time contested Native American territory. Although very incomplete, Cope immediately recognized the importance of these specimens as a very primitive theropod dinosaur.

Coelophysis lived during the Late Triassic Period and is about 200 million years old; its remains are known from the Southwestern United States. Several other animals have been named that are very similar to *Coelophysis*. Among these, *Rioarribasaurus*, *Longosaurus* and *Megapnosaurus*, all discovered in close geographical proximity to the original *Coelophysis* finds, are now considered to be the same species as *Coelophysis*. Others, like *Syntarsus* from Zimbabwe, are known from finds thousands of miles away. Yet this can easily be explained by the fact that during the time *Coelophysis* lived, the continents were not arranged as they are now: they were still united into a single supercontinent, Pangea, so it is not surprising that the entire area comprised

a similar fauna as continents with large land masses do today. Because the name *Syntarsus* is preoccupied by a beetle, many paleontologists now feel the name should be replaced by *Megapnosaurus*.

Besides being primitive, *Coelophysis* is noteworthy for other reasons. In 1947, expeditions by the American Museum of

> This representation of *Coelophysis* portrays it as a snake-necked carnivore. Stomach contents show that it fed on small reptiles.

∨ In 1947 Museum paleontologists Edwin Colbert and George Whitaker discovered the *Coelophysis* quarry at Ghost Ranch, New Mexico.

Natural History to Northwestern New Mexico discovered an exceptional accumulation of *Coelophysis* specimens. These were found at a place that goes by the colourful name Ghost Ranch, which at the time was a tourist destination for "greenhorns" who wanted to try their hand at horse riding and camping. It was also the seasonal home of artist Georgia O'Keefe, who was a frequent visitor to the excavations, and many of her paintings portray the landscape and the exact rocks where the specimens were found. What the Museum field parties found was a mass accumulation of *Coelophysis* specimens ranging from small animals to adults. As *Coelophysis* was not a huge animal – the adults were only about 3m (10ft) in length and most of this was neck and tail – large numbers of individuals could be excavated in single blocks.

Almost all of the animals found in the Ghost Ranch quarries are remains of *Coelophysis*. A few other species have also been found, however, including the remains of terrestrial crocodile-line archosaurs, fishes, varied unusual reptiles called drepanosaurs and non-dinosaurian dinosauromorphs. The circumstances behind these accumulations have been difficult to determine. The fossils are usually found in single bedding planes

indicative of a single event; however, it was probably not a catastrophic one. According to the most recent interpretations, the Ghost Ranch deposits represent ephemeral ponds, much like the seasonal watering holes that occur in East Africa today. During the rainy season these attract vast numbers of animals, which when the seasons change sometimes succumb to dehydration. Perhaps an aspect of their behaviour facilitated this preservation. This is further suggested by the fact that the closely related *Syntarsus* has been found in exactly the same sort of bone beds in southern Africa. One problem with this explanation is that it does not explain why most of the animals found at the deposit were the carnivorous *Coelophysis*. In ecosystems today, herbivores vastly outnumber carnivores. There are far more deer, antelope, rabbits and gazelles than there are wolves, cheetahs, snow leopards and coyotes. Yet Ghost Ranch is not unusual in this regard, as several dinosaur-dominated communities show this same pattern, where a much higher percentage of carnivorous animals are found relative to herbivores.

The number and quality of preservation allowed a good deal of early paleobiology to be done on the specimens found in the

Ghost Ranch quarry. Unlike nearly all dinosaur deposits, most of the specimens were completely articulated. Many even showed their necks pulled back, suggesting that there may have been some desiccation before burial. A couple of the specimens even showed remnants of their last meal in the abdominal cavity. The first interpretation of these remains is that they are the bones of small juvenile *Coelophysis*. Consequently, *Coelophysis* was given the reputation as a cannibal dinosaur, consuming its own young in what was, at the time, considered to be a monospecific assemblage.

The main specimen in question is on display in the Hall of Saurischian Dinosaurs at the Museum. It is such a beautiful example that a bronze cast of it was made to adorn the wall of the downtown B and C train stop at the 81st Street–Museum subway station. One night a graduate student was leaving the Museum after work. Thoroughly acquainted with the anatomy of both *Coelophysis* and other Late Triassic archosaurs, he spent time looking at the bronze *Coelophysis* on the wall while waiting for the train. His immediate thought was, "Those don't look like *Coelophysis* bones." Over the next few days, after looking

at the specimen on display, he confirmed his original observation. Following careful comparison, he was able to show both on the basis of general shape, and with the assistance of a small team, that the architecture of the bones indicated that these were not the bones of juvenile or baby *Coelophysis*. Instead they were the remains of crocodile-line archosaurs. Instead of the fearsome cannibal midget dinosaurs popularized in the book *The Little Dinosaurs of Ghost Ranch*, by Museum curator Ned Colbert, *Coelophysis* was an active, opportunistic predator.

∧ The Late Triassic *Coelophysis* bonebed at Ghost Ranch New Mexico preserves hundreds of animals.

⊓ Early on it had been proposed that *Coelophysis* was a cannibal due to bones found in its abdominal cavity. Subsequent research has shown that these were the remains of small crocodile-line animals.

> The intimate association between many of the specimens at Ghost Ranch indicates that there was a mass death event.

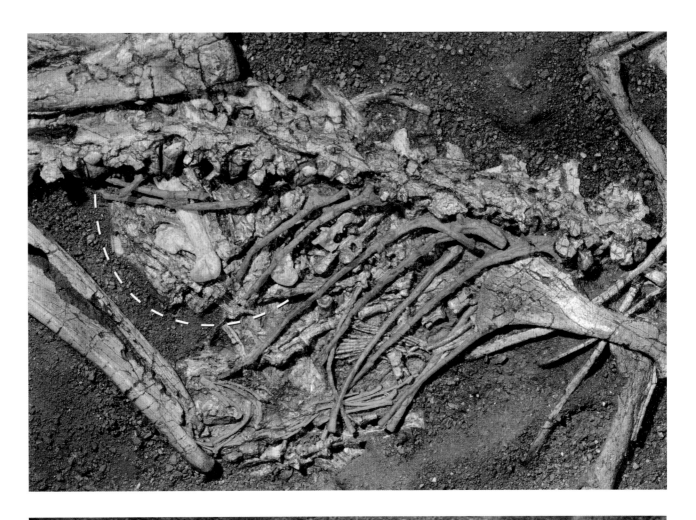

DILOPHOSAURUS WETHERILLI

EARLY JURASSIC
KAYENTA FORMATION
NORTH AMERICA

ONE OF THE MEMORABLE VILLAINS OF THE MOVIE *JURASSIC PARK*, *DILOPHOSAURUS* WAS PORTRAYED AS A FRILLED, POISON-SPITTING CARNIVORE. IS THERE ANY EVIDENCE FOR THIS? NO. *DILOPHOSAURUS* WAS A 6M (19½FT) LONG, BIPEDAL DINOSAUR THAT LIVED IN AN ARID AREA OF WHAT IS NOW NORTHERN ARIZONA ABOUT 196–183 MILLION YEARS AGO.

*D*ilophosaurus is an important dinosaur for many reasons. It fills an important intermediate phylogenetic position between more primitive theropod dinosaurs, such as coelophysids, and the advanced tetanurans. It is also notable in that it has two crests on its head, hence its name that translates as "two-crested lizard". Head crests are ubiquitous features of dinosaurs in general, and many theropods in particular, and evolved independently in several groups of dinosaurs.

In many species, crests are found on the top of the muzzle. One thing that most crests had in common is that they were very thin – so thin, in fact, that it is difficult to understand how they could have survived if the animal was an active carnivore. The crests of *Dilophosaurus* and the tyrannosaur *Guanlong* are paper-thin. Several functions have been suggested, which range from thermoregulation (now discredited), to species recognition and sexual selection. Because all of the known specimens with skulls have crests, it could be argued that sexual selection is not a viable explanation.

However, in some living animals (like rhinoceros, antelope, and elephant), tusks and horns are present in both sexes, albeit of different sizes. The larger structures in males are thought to be ornaments to attract mates: this is called mutual sexual selection. Unfortunately, we do not have a large enough sample of *Dilophosaurus* specimens to test this hypothesis (or of almost any other dinosaur).

Dilophosaurus is also one of those rare dinosaurs for which we have evidence of trackways. While dinosaur tracks are very common across the globe, it is often very difficult to assign specific tracks to a particular dinosaur species. The case for *Dilophosaurus* is a good one, but not 100 per cent definitive. A few decades after the original discovery on the Navajo Nation reservation, paleontologists found a significant number of dinosaur tracks in

> *Dilophosaurus* depicted laying down. This was probably not its regular resting posture; rather, it would have crouched like a chicken.

almost the same beds where the *Dilophosaurus* material was collected. Some have suggested that these trackways indicate that these animals travelled in groups, while others dispute this. What is apparent in these tracks, however, is something extremely unusual.

In the presumed Dilophosaurus trackways, there are a few furrows that show a dragging tail. This is rarely found in any other dinosaur trackways, as all evidence indicates that the tails were held parallel to the ground, supported by stiff tendons attached to the pelvis. The furrows are not continuous, so there is no evidence that they dragged their tails "Godzilla"-style, but rather in a few individuals the tails did come into contact with sediments. These trackways also preserve something that is rarer still, and that is the preservation of an animal resting.

It is one of the few instances where there is evidence of imprints of the hands. The resting depression is on an incline with the head oriented uphill. This may suggest that it was easier for the animal to stand up on an incline than on flat ground, though there is no evidence that it used its forelimbs to prop itself up. A footprint right in the middle of the body depression is evidence that the animal moved off in a healthy, sprightly fashion.

∨ Many trackways that are nearly coeval with the *Dilophosaurus* bones have been ascribed to this dinosaur. However, tracks are difficult to interpret. What we can determine is that these sorts of animals were common in their day.

> A computer reconstruction of *Dilophosaurus* depicting it sprinting.

↘ A *Dilophosaurus* specimen as it was found with the classic death pose with head pulled back.

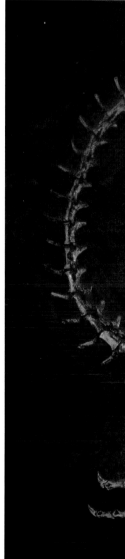

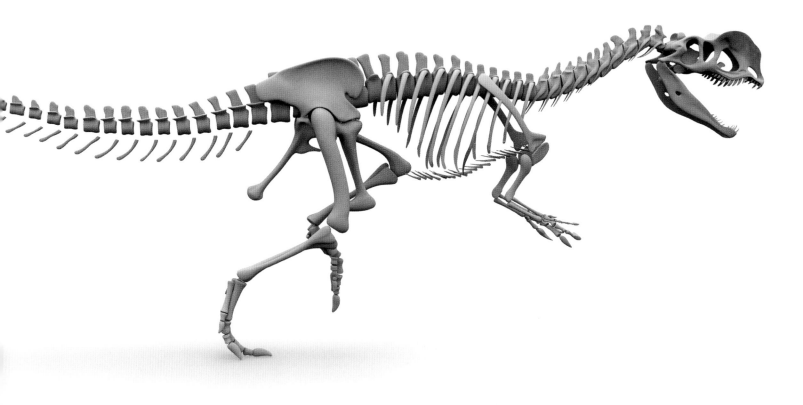

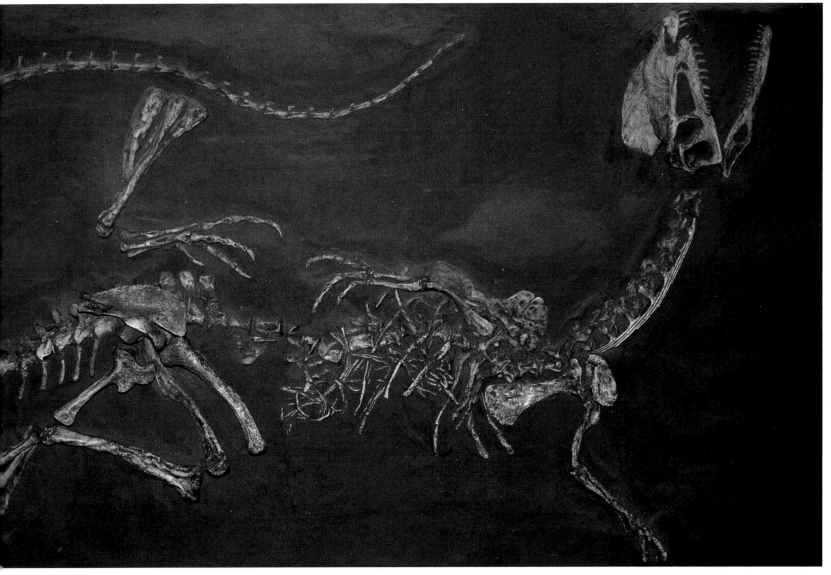

ALLOSAURUS FRAGILIS

LATE JURASSIC
MORRISON FORMATION
WESTERN NORTH AMERICA

THE FIRST OF THE WELL-KNOWN BIPEDAL MEGA CARNIVORES WAS *ALLOSAURUS*. A MEDIUM-SIZED THEROPOD, ABOUT 8M (26¼FT) LONG, *ALLOSAURUS* LIVED ABOUT 155 MILLION YEARS AGO IN WHAT IS NOW WEST-CENTRAL NORTH AMERICA. ALONG WITH *TORVOSAURUS* AND *CERATOSAURUS*, IT FILLED THE APEX PREDATOR POSITION IN THE LOCAL ECOSYSTEMS.

Allosaurus is known from literally hundreds of specimens. The early ones found by O.C. Marsh and E.D. Cope's field parties were very fragmentary. However, one specimen was excavated by Cope's collectors in 1877 at the legendary Como Bluff site in Wyoming, and this was purchased together with the body of Cope's fossil collection by the American Museum of Natural History in the 1890s. Cope had never unpacked the specimen, and when it was opened and studied, it turned out to be the most complete large theropod dinosaur known at that time. By 1898 it had been mounted for public display, and this mount is notable in that museum president Henry Fairfield Osborn had it articulated in a lifelike action pose feeding on the carcass of an *Apatosaurus*. This scene inspired a well-known painting by Charles R. Knight, depicting the scene 155 million years ago. This skeleton has never been studied in depth. Since then many new *Allosaurus* skeletons have been collected and several of these are virtually complete. Closely allied animals have been found in similarly aged rocks in Portugal and possibly Tanzania.

One remarkable site has produced thousands of *Allosaurus* bones representing several different age classes. Known as the Cleveland Lloyd Dinosaur Quarry, it lies south of Provo, Utah. While other species have also been found at the site, it is unusual in that there are so many bones from a carnivorous species – usually the skeletons of carnivorous animals are far less common than the remains of their prey. It has been suggested that the area that would become the Cleveland Lloyd Dinosaur Quarry was a predator trap around a water hole. Here *Allosaurus* may have become mired in sticky, muddy sediment, only for their skeletons to decay and become disarticulated. Although some have suggested that *Allosaurus* may have hunted communally, there is little if no actual fossil evidence in support of this.

There is a great deal of evidence suggesting that *Allosaurus* was an active

> *Allosaurus* holds a commanding presence, both in the history of dinosaur discovery and its impact on dinosaur science. This is a juvenile.

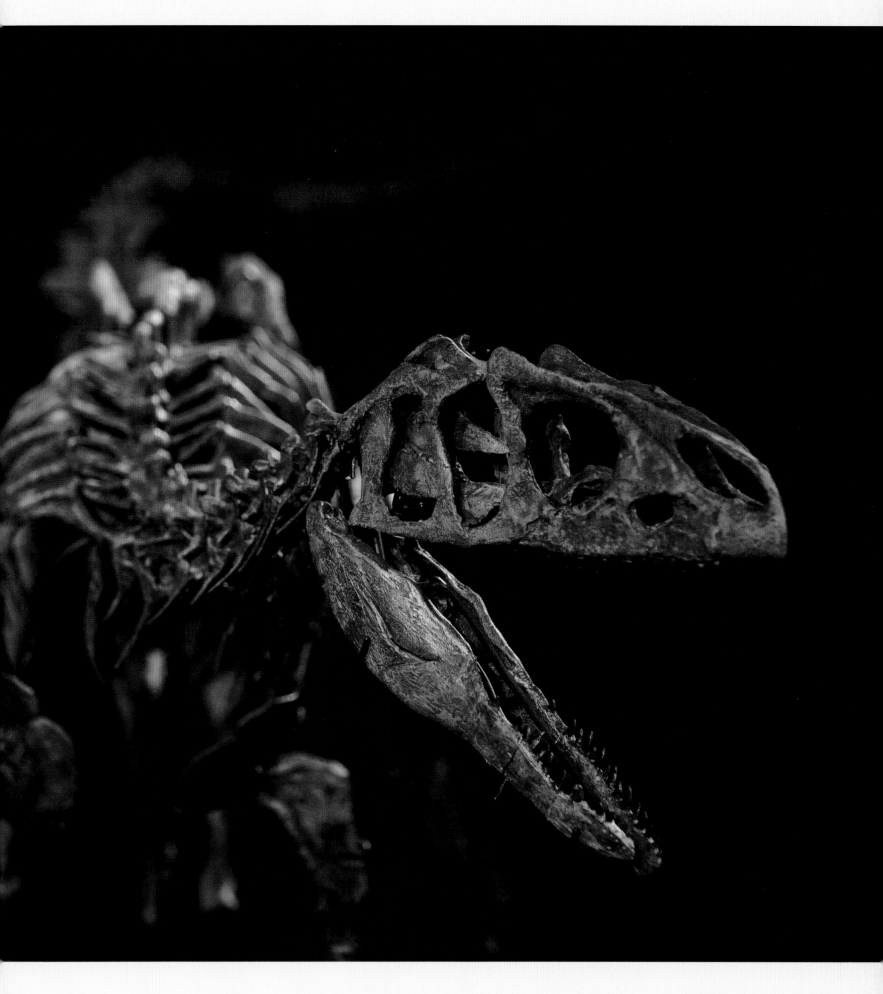

egg-laying females of modern birds, has been reported for one specimen, and while it could be that this may allow us to determine males from females, this evidence has been contested.

The specimens vary in size and it has been determined that in young animals the legs were proportionally longer – just as puppies grow into their disproportionately big feet, large carnivorous dinosaurs grew into their legs. This also probably indicates that the young had different hunting strategies to the adults. This occurs in some animals today: Komodo dragons, the largest living lizards (up to 3m/10ft) in length) have hatchlings that are only 50cm (19½in) long. The young are largely arboreal and feed on insects and other small animals, while the adults are capable of killing large pigs and even humans. Like Komodo dragons, *Allosaurus* is likely to have changed its diet, habitat, and hunting habits as it grew. There is possible evidence of cannibal behaviour, as juvenile *Allosaurus* teeth have been collected near the ribs of reasonably complete *Allosaurus* adults. Finally, the *Allosaurus* brain was much less birdlike than more advanced theropods, and it had only limited binocular vision. This further suggests that its social system may not have been as sophisticated as the more advanced dinosaurs that occurred closer to the origin of birds.

predator. For example, there is an example of an *Allosaurus* vertebra that may have a puncture wound caused by the flailing tail of a stegosaur. And as is typical of large theropods such as *Tyrannosaurus*, many *Allosaurus* specimens show signs of a tough life – multiple healed fractures occur in several specimens, and stress fractures are even more common. One of these is nicknamed "Big Al"; 19 of its bones were either broken or showed signs of bone infections. These injuries may have been the result of inter-species battles for territory or mates or the consequences of predation gone wrong.

Because of the number of specimens that are known, we understand a great deal about the paleobiology of *Allosaurus*. It was the dominant predator in Late Jurassic rocks of what is now North America; other large predators are also known from these sediments, but these occur with far less frequency. We know that *Allosaurus* stopped growing between the ages of 20 and 28. Medullary bone, which occurs only in

⌐ This *Allosaurus* mount at the Museum depicts the animal feasting on a carcass. Completed in 1915, it was the first time a dinosaur had been mounted in an active, lifelike scenario.

< This excellent painting of *Allosaurus* by Charles R. Knight, was a study of the eventual mount.

∧ The craftsmanship that went into the early mounts of fossil animals is an aesthetic that is difficult to replicate today.

> The Cleveland Lloyd Quarry south of Provo, Utah is an *Allosaurus* bonebed where innumerable individuals are preserved.

ALBERTOSAURUS SARCOPHAGUS

LATE CRETACEOUS
HORSESHOE CANYON FORMATION AND OTHER LOCATIONS
WESTERN NORTH AMERICA

ALBERTOSAURUS WAS THE APEX CARNIVORE OF MUCH OF NORTH AMERICA BEFORE THE APPEARANCE OF *TYRANNOSAURUS REX*. IT IS CLOSELY RELATED TO *TYRANNOSAURUS REX*, BUT MORE LITHE IN APPEARANCE THAN ITS LARGER RELATIVE. IT GREW TO A MAXIMUM SIZE OF AROUND 10M (32¾FT) AND 2 TONNES.

Recent decades have seen the discovery of many new species of tyrannosaurs in the northern hemisphere. Nearly all are from North America, but some have also been found in Asia. This has much to do with the geography of the northern hemisphere at the time these animals lived: what we think of as North America today was divided into two provinces, a western one and an eastern one, separated by a vast, shallow inland sea that extended from today's Gulf of Mexico as far as Canada. This sea was populated by a diverse group of creatures, including large, seagoing reptiles like mosasaurs and plesiosaurs, together with giant turtles and immense predatory fish; toothed flightless birds chased fish in these waters and giant pterosaurs filled the skies. Later in the Cretaceous, sea levels began to fall, and many believe that large tyrannosaurs emigrated out of western North America to the eastern provinces and across the

Bering Strait into what is now modern Asia. In both of these areas, large tyrannosaurs that are closely related to *Tyrannosaurus rex* appear. In eastern North America this was *Appalachiosaurus*, with *Zhuchengtyrannus* and *Tarbosaurus* in Asia. The southern part of North America also appears to have been a tyrannosaur incubator, where animals such as *Lythronax*, *Bistahieversor* and *Teratophoneus* have been found.

Many specimens of *Albertosaurus* have been discovered, and there could be more as there has been some dispute and confusion over whether another named species, *Gorgosaurus*, is in fact *Albertosaurus*, rather than a species in its own right. After extensive study, however, it appears that specimens

∧ *Albertosaurus* was a small, more lightly built version of *Tyrannosaurus*. It also lived a little earlier in time, yet it is still very well represented in the fossil record.

⅂ The fossil hunter Charles H. Sternberg in the preliminary stages of the excavation of a juvenile *Albertosaurus*. This specimen is now on display in the Museum saurischian dinosaur hall.

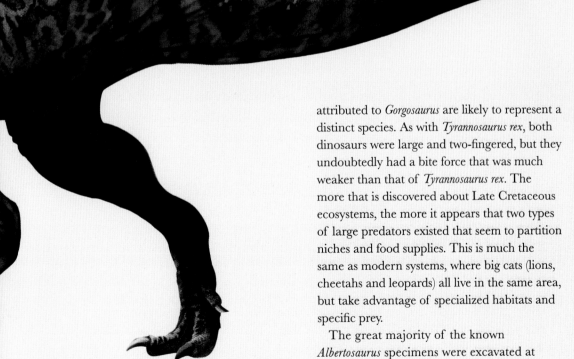

attributed to *Gorgosaurus* are likely to represent a distinct species. As with *Tyrannosaurus rex*, both dinosaurs were large and two-fingered, but they undoubtedly had a bite force that was much weaker than that of *Tyrannosaurus rex*. The more that is discovered about Late Cretaceous ecosystems, the more it appears that two types of large predators existed that seem to partition niches and food supplies. This is much the same as modern systems, where big cats (lions, cheetahs and leopards) all live in the same area, but take advantage of specialized habitats and specific prey.

The great majority of the known *Albertosaurus* specimens were excavated at one site, called the Dry Island Bone Bed.

It was discovered in Alberta, Canada, in 1910 by American Museum of Natural History paleontologist Barnum Brown. These expeditions to Alberta, along the Red Deer River, were novel in the sense that the intrepid paleontologists lived on a flatboat and endured harsh, mosquito-infested conditions. They paddled and motored the boat down the river, stopping and mooring every few days to carry out their excavations. This locality was later rediscovered in 1997 by Canadian paleontologists using Brown's photographs and notes, and excavations continued until 2005. Not only did they find some of Brown's collecting artefacts, they even located parts of some of the same

specimens held in the Museum collections. In all, about 26 individuals were found in this quarry, providing a great representative population sample.

Detailed analyses of these bones showed that an age distribution, or growth curve, could be constructed for these individuals. This told us that very few of these animals had reached adult size, which, as for other advanced tyrannosaurs, occurred at about 17 years of age. The oldest individual in the group was 28 years old and 8.5m (28ft) in length, the smallest was two years old and weighed about 50kg (110lb). The average age of *Albertosaurus* specimens that have been analyzed is 14 years. Although *Albertosaurus* didn't grow as fast as *Tyrannosaurus*, it did grow quickly, which means that the behaviour, habits and diet of subadults was probably not unlike those of fully grown individuals.

∧ This is the old way to look at dinosaurs. While the tail is not dragging on the ground, the entire front part of the skeleton should fall forward so the spine is parallel with the ground.

∧ Many exceptional specimens of *Albertosaurus* have been found, including several three-dimensional skulls.

∨ Conditions in Alberta for Barnum Brown's crew were tough. Access to exposures was limited so they lived on a flatboat. According to their notes, they were almost eaten alive by mosquitos, black flies, and deer flies.

TYRANNOSAURUS REX

LATE CRETACEOUS
HELL CREEK FORMATION
WESTERN NORTH AMERICA

IF I DID A WORD CHECK ON THIS BOOK, *TYRANNOSAURUS REX* WOULD GET THE MOST MENTIONS WHEN IT COMES TO A SPECIES OF DINOSAUR. IF A PERSON KNOWS A SINGLE DINOSAUR NAME, IT IS APT TO BE *TYRANNOSAURUS*.

The first recognized *Tyrannosaurus rex* specimen was excavated in 1902 by Barnum Brown in eastern Montana. It was given the name *Tyrannosaurus rex* by Henry Fairfield Osborn in 1905. However, numerous fragmentary fossils of what would come to be known as *Tyrannosaurus* had been found decades earlier from the same general area, but not recognized as particularly special.

The original specimen was not complete, yet the magnitude and importance of this animal was immediately apparent, as expressed in Brown's letter to Osborn at the time: "Quarry No. 1 contains the femur, pubes, part of the humerus, three vertebrae and two indeterminate bones of a large Carnivorous Dinosaur, not described by Marsh. … I have never seen anything like it from the Cretaceous."

Three years later in 1905, Brown found another specimen that was much better preserved. This is the iconic *T. rex* specimen in the dinosaur halls of the American Museum of Natural History. When this specimen was mounted at the Museum in 1915, it created quite the sensation, and people lined up in droves to view this giant prehistoric carnivore. According to Brown, the original specimen was sold to the Carnegie Museum in Pittsburgh in 1941. It was wartime and Brown stated that this was because "we were afraid that the Germans might bomb the American Museum in New York as a war measure, and we hoped that at least one specimen would be preserved."

Very little was known about dinosaur posture and biomechanics when the iconic specimen was mounted. When originally conceived by Osborn, he was going to mount two *Tyrannosaurus* together in a snarling duel over a carcass of a deceased

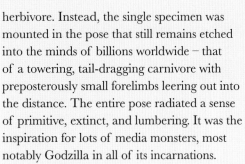

< As relatives of
Tyrannosaurus have been found
with a covering of simple
feathers, it is possible
that it may have had them on
portions of its body as well.

∟ The Museum's *Tyrannosaurus*
skeleton was excavated in the
Hell Creek badlands of Montana
in 1905.

herbivore. Instead, the single specimen was mounted in the pose that still remains etched into the minds of billions worldwide – that of a towering, tail-dragging carnivore with preposterously small forelimbs leering out into the distance. The entire pose radiated a sense of primitive, extinct, and lumbering. It was the inspiration for lots of media monsters, most notably Godzilla in all of its incarnations.

In 1995 the Museum opened its renovated dinosaur halls. The highlight of the saurischian dinosaur hall was the remounted *T. rex*. Instead of Godzilla, the new mount was posed in a way that looked like it was stalking you. This new pose was dramatic in that the tail was held aloft, parallel to the backbone, and the head was supported by an S-shaped neck. The new pose looked more like a giant-toothed pigeon than a reptilian kangaroo.

In 2010 my colleagues and I authored a paper in the prestigious journal *Science* called "Tyrannosaur Paleobiology: New Research on Ancient Exemplar Organisms". Basically, it laid out the general outline of what we knew then about *T. rex* and its close relatives. Our agenda was to use the charisma of *T. rex* to talk about what is generally being done in dinosaur science. The paper said a lot and focused on all aspects of tyrannosaur biology. But that was then, this is now. Work on all aspects of this magnificent animal has since been honed and refined. New technologies have come into play and more specimens have been found. Let's just examine a few.

If I was writing about a living animal, I would start with some specifics. The weight of *T. rex* varied between 5,900kg (13,000lb) and a top end estimate of 14,500kg (32,000lb). Wow! This is a variance of over 2.5 times. This speaks loudly to how difficult it is to empirically measure how much extinct animals weighed. Most reasonable estimates are in the high 8,400kg (13,000lb) approaching the 14,000kg (30,000lb) level.

Physical measurements – tooth length, leg length, brain size – are easy to quantify. Information such as growth bite force, hunting strategy, or telling males from females is more difficult. Let's start with reproduction. Making an egg requires a huge amount of calcium mobilization. Calcium is hormonally regulated, even in our bodies, and is the reason that women are much more susceptible to osteoporosis than men. In many archosaurs, including *T. rex*, a specific kind of bone may have been found. This bone type is called medullary bone. Medullary bone serves as a calcium reservoir for building eggshell. It is found in living female birds and has also reportedly been found in *T. rex* specimens,

allowing us to suggest that some of the specimens may have been gravid females.

Moving towards physical properties, the speed of an adult *T. rex* has been recalculated. Previously it had been suggested that it could move in the range of 20 miles an hour (32 kilometres per hour). Using new computer models, we can calculate that it was much more of a sloth at 12mph (19kph), which the fittest of us could outrun.

Other interesting facts are that it was one of the animals which saw the asteroid at the end of the Cretaceous period, and although it was giant, its brain size was proportional to its body weight compared to other closely related dinosaurs. We also know from damage to the

∧ Rock overlying the *Tyrannosaurus* bones in the quarry was removed using a team and shovel by Barnum Brown and field crew.

> A head-on view of the imposing skull of *Tyrannosaurus*. Its eyes probably faced more forward than those of other large carnivorous dinosaurs, giving it better binocular vision.

skulls, that they (presumably the males) bashed their heads and bit each other. Presumably, this was, like in many living animals, done in competition for territory, food or mates.

Body covering for *Tyrannosaurus rex* has until recently been the subject of much conjecture. *T. rex* has traditionally been depicted as fairly grotesque and reptilian, with large scales, and sometimes horns and spikes. New specimens of *Tyrannosaurus rex* and its close relatives give direct evidential support to what it looked like. These include specimens of *Yutyrannus* (see p. 17) which are completely cloaked in feathers. Small patches of preserved skin have been found associated with newly collected *T. rex* specimens. These scales are actually small raised tubercles, only about 3mm in diameter. We do not have skin from all over the body, so we don't know if tubercle size was heterogenous. Nevertheless, with the evidence at hand, we would predict that *T. rex* was less grotesque, covered by an almost velvety fabric of supple scales, with small primitive feathers sparsely planted over the body.

As we indicated in our review article mentioned above, we probably know more about *T. rex* than any other dinosaur. And there are 50 specimens known of this amazing animal. It continues to be a subject of fascination, a popular icon, and probably the first dinosaur name imprinted in the minds of children globally. Besides all this, it is the inspiration for budding paleontologists worldwide.

Long may the king reign.

< Mounted in an active stalking pose, the T. rex skeleton is a centrepiece of the Museum's Hall of Saurischian Dinosaurs.

STRUTHIOMIMUS ALTUS

LATE CRETACEOUS
OLDMAN FORMATION
NORTH AMERICA

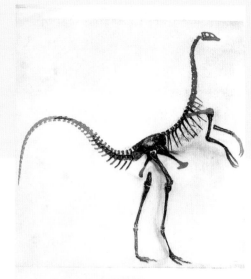

EARLY ON IT WAS NOTICED THAT MANY ADAPTATIONS SEEN IN NON-AVIAN DINOSAURS PRESAGED MORPHOLOGIES SEEN IN MODERN BIRDS IN A CONVERGENT WAY. OSTRICH DINOSAURS (ORNITHOMIMIDS) ARE A GOOD EXAMPLE OF THIS. ONE OF THE BEST-KNOWN OSTRICH DINOSAURS IS *STRUTHIOMIMUS*.

It lived about 78 million years ago in what is now western North America and was about 3m (10ft) in length. Ornithomimidae, the group to which it belongs, was primarily Laurasian or northern hemisphere in its distribution. Ostrich dinosaurs were never a very diverse group, but they do display many interesting features that evolved independently in modern birds. *Struthiomimus* means "ostrich mimic" and one look at the skeleton affirms why it was given this name. Like an ostrich, it had a long neck, large eyes, and long hind limbs, and there is direct fossil evidence that it had a keratinous beak like a modern bird. Some have estimated that it could reach speeds of up to 50mph (80kph), which is faster than an African ostrich, but no rigorous biomechanical studies have been conducted.

Over the last few years it has been demonstrated that primitive members of the group had teeth. In one species, *Pelecanimimus* ("pelican mimic") from the Early Cretaceous of Spain, the teeth were small and closely packed into the jaws. As the name implies, there is a great deal of soft tissue in the throat region, which may have been the remnants of a pouch as seen in modern pelicans. Other primitive ornithomimids, such as *Harpymimus* and *Shenzhousaurus*, had a surprising tooth configuration, with just a few widely spaced teeth restricted to the lower jaw. Other ornithomimids are toothless. This diversity of dentition and beak shape

∧ A *Struthiomimus* specimen mounted. It is immediately obvious why these are colloquially called ostrich dinosaurs.

< A beautiful vintage chalk drawing by Museum illustrator Erwin Christman from the early 1900s. This shows that even early on in dinosaur studies the animals were sometimes portrayed as active animals with their tails aloft.

⌐ Two excellent specimens of another "ostrich dinosaur", *Gallimimus*, that thankfully were repatriated to Mongolia following their illegal export by fossil poachers.

indicates that members of this group may have had very diverse diets.

The morphology of the hands, especially in the more advanced forms, is unusual in that all three of its fingers were about the same length and terminated in long, straight claws. This, coupled with the fact that the claws on the feet were also not strongly recurved, suggests that these animals were not active hunters.

This view (at least for the Late Cretaceous Mongolian form *Gallimimus*) is further supported by the presence of soft tissue on the beak. This specimen preserves a comb-like structure reminiscent of the beak of ducks. Modern ducks sieve out food from the water using this structure, and the comb may have had a similar function in ornithomimids.

Until recently the body covering of ornithomimids was unknown. In the last couple of decades specimens of *Ornithomimus* from the Late Cretaceous of Canada have been found which preserve carbonized traces of what have been interpreted as pennaceous feathers. These are particularly abundant along the forelimbs, and it has been suggested that these feathers were used in display as in modern ostriches. Juveniles with feathers from these same beds preserve only plumulaceous feathers, thereby supporting the idea that the pennaceous feathers were only developed in adults as display structures.

Images of more primitive ornithomimids from the Early Cretaceous of China circulate on the internet. These specimens of animals like *Shenzhousaurus* show a complete body covering of plumulaceous feathers. Sadly, these specimens have not made their way into public scientific collections and have disappeared into the burgeoning collectors' market.

DEINOCHEIRUS MIRIFICUS

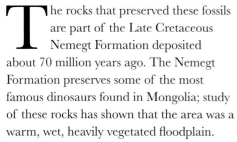

ONE OF THE MOST SPECTACULAR AND UNEXPECTED FINDS OF THE POLISH-MONGOLIAN EXPEDITIONS OF THE 1960S AND EARLY 1970S WAS AN ENORMOUS SET OF ARMS. THESE WERE FOUND AT THE ALTAN ULA LOCALITY BY FAMED POLISH PALEONTOLOGIST ZOFIA KIELAN-JAWOROWSKA IN SOUTHERN MONGOLIA IN 1965.

The rocks that preserved these fossils are part of the Late Cretaceous Nemegt Formation deposited about 70 million years ago. The Nemegt Formation preserves some of the most famous dinosaurs found in Mongolia; study of these rocks has shown that the area was a warm, wet, heavily vegetated floodplain.

This set of arms was nearly 2.75m (9ft) long. Only the forelimbs and part of the shoulder girdle were found. When another Polish paleontologist, Halszka Osmólska, originally described the specimen, she did not have much to work with. She named it *Deinocheirus*, which means "terrible claw" – a fitting name, since a single talon measures 20cm (8in) long and is perched on a massive hand. Originally it was thought to be a member of what was then known as "Carnosauria" – a group of large carnivorous dinosaurs that included many well-known animals such as *Tyrannosaurus* and *Allosaurus*. This group is no longer recognized as valid.

Comparisons were also made initially with

therizinosaurs whose remains have also been found in Mongolia. The first therizinosaur fossil, *Therizinosaurus*, was found in the same fossil beds in which *Deinocheirus* was discovered. Early on, therizinosaurs were poorly known and turned out to be among the strangest of all dinosaurs. They had immense bodies supporting tiny heads, short tails, and huge claws. One of these from *Therizinosaurus* itself is over 30cm (11¾in), and this bony part of the claw would have supported a keratinous claw sheath (like a human fingernail) that was at least 60 per cent larger. As fearsome as these beasts appeared, they were plant eaters and probably used their long claws to pull branches down into their jaws, much in the same way that sloths do today.

Over the years, scientists began to establish that *Deinocheirus* was neither a "carnosaur" nor a therizinosaur. Instead it was determined to be an ornithomimid, and a giant one at that. Because ornithomimids have such long arms, *Deinocheirus* is not as large as it may seem. If the proportions

were scaled up from other ornithomimids, *Deinocheirus* would have had a length of about 10.7m (35ft), which is big but not necessarily huge. Enormous sizes were originally postulated because people were extrapolating from other theropods with much smaller forelimbs.

Our understanding of many specifics of *Deinocheirus* and its relationships has changed dramatically over recent years. This has been fuelled by the discovery of several new specimens. Some of these were excavated at the original *Deinocheirus* locality by a Canadian–Mongolian team, and another was found at the Bugin Tsav locality, which is not far from the Altan Ula beds. Some of these specimens showed clear signs of having been looted; the looting of dinosaur specimens by commercial interests is a major problem, not just in Mongolia but throughout the world.

It became known that the specimen had been trafficked out of Mongolia and sold to Europeans through a Japanese middleman. Fortunately, it was intercepted by a

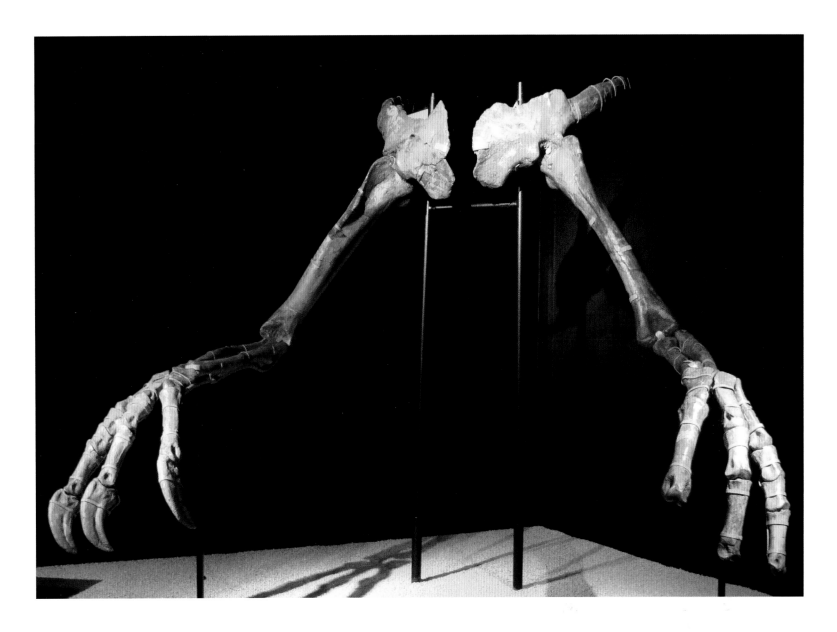

legitimate European fossil dealer, who arranged for the specimen to be deposited in a museum in Belgium. From there it was repatriated to Mongolia, where it was combined with the rest of the skeleton. This fantastic new material gives us a remarkable picture of this very unusual animal.

The new specimen clearly supports the ornithomimid identity of *Deinocheirus*. The most striking thing about its appearance, however, is that it sported a large fin on its back supported by extensions of the backbone segments. Fans like this occur in a broad variety of unrelated dinosaurs and there is little consensus on what they were used for; everything from heat regulation to – believe it or not – "wind sails" has been proposed. It is most likely they were used for display in a way similar to many of the unusual structures found on living animals. It

also had bulky feet, suggesting slow, lumbering movement – rather sloth-like.

Skulls tell us a lot about animals: *Deinocheirus* had a smaller than average brain for its size and may have had a better sense of smell than many dinosaurs. A study of its skull biomechanics indicates that it had a weak bite force and may have simply slurped its food from the bottom of ponds and lakes. This isn't as odd as it seems: another ornithomimid, a specimen of *Gallimimus* collected by American Museum of Natural History–Mongolian Academy of Sciences expeditions in 2000, showed the presence of keratinous filters on the beak, similar to those ducks have today. Associated with the skeleton were nearly 1,500 gastroliths, or small stones that formed a gastric mill in the stomachs of the animal. Such stones have also been found in the

primitive toothed ornithomimid *Shenzhousaurus*.

A cast of the grasping arms of the type specimen hangs in the Hall of Saurischian Dinosaurs at the Museum and provides a frequent frame for many a selfie. They were rather mysterious until recently. Now we know that, although unusual, *Deinocheirus* fits nicely within the paradigm of dinosaur biology developed over the last two decades.

⌐ A big head, huge claws, humped back and a feathery body covering make *Deinocheirus* look like a Dr. Seuss animal.

∧ When it was first discovered, paleontologists had no idea what kind of animal *Deinocheirus* was. Some suggested it was huge, larger than any other theropod and equally grotesque.

COMPSOGNATHUS LONGIPES

JURASSIC
SOLNHOFEN FORMATION
GERMANY

COMPSOGNATHUS IS A POORLY KNOWN BUT IMPORTANT SMALL DINOSAUR FOUND IN LAGOONAL SEDIMENTS IN THE SAME FORMATION AS *ARCHAEOPTERYX*. IMMEDIATELY IT WAS RECOGNIZED THAT THE ANATOMY OF *COMPSOGNATHUS* WAS VERY SIMILAR TO THE PRIMITIVE "BIRD" *ARCHAEOPTERYX*. IT WAS SO SIMILAR THAT IT WAS THE FIRST NON-AVIAN DINOSAUR TO BE RECONSTRUCTED WITH FEATHERS BY THOMAS HUXLEY IN 1876.

Perhaps because *Compsognathus* was so apparently birdlike, and was found so early in the history of dinosaur study – 1859 – a number of strange theories about it have been proposed. One is that it used its forelimbs (which were thought at the time to have only two fingers, but actually have three), to propel itself through water using its arms as oars like a steamer duck. All of the evidence now suggests that these animals were fairly traditional bipeds with a conservative body plan. They probably lived along the shores of a shallow sea. *Compsognathus* is one of the few dinosaurs for which we have a very good idea what they ate: both of the known specimens have the remains of small lizards in their body cavities, providing good evidence that these animals were lacertophages.

While *Compsognathus* was a small animal, there are much larger individuals from other closely related species. *Huaxiagnathus* from the Early Cretaceous of Northeast China is much larger, approaching 2m (6½ft). Another compsognathid is *Sinosauropteryx*.

Sinosauropteryx is a revolutionary animal; although it had long been predicted, it was the first non-avian dinosaur to be collected that preserved indisputable evidence of feathers. This caused quite a stir in the late 1990s and of course there were doubters. It will suffice to say here that after intensive study, the original claims of a feathered animal are borne out. As an aside, the first specimen of *Sinosauropteryx* (like *Compsognathus*) also preserved a lizard in the body cavity.

The feathers of *Sinosauropteryx* are not like the feathers of a modern bird. Instead of long branching structures, which are very heterogeneous across the body, all of the *Sinosauropteryx* feathers were 13–50mm (½–2in) in length and uniform in structure over the entire animal. They were short,

> A fluffy *Compsognathus* reconstruction. It is known from stomach contents that it fed on lizards, and its diet probably included insects.

tubular bristles with the longest ones just above the shoulders. *Sinosauropteryx* is a fairly primitive theropod, so it is not unexpected to find primitive, undifferentiated feathers. On the tail of one of the specimens you can observe variations in colour, which indicates that in life the tail was banded. Most paleontologists think that these feathers, and feathers in general, originally evolved as thermal blankets to insulate these animals from the environment.

The original (and only) specimens of *Compsognathus* do not feature traces of feathers, but this does not mean they were not present. Some of the *Archaeopteryx* specimens from these same localities do not preserve feathers, even though most specimens do. In addition, there are other small dinosaurs from these beds that display surprising integumentary structures. Two of these are *Sciurumimus* and *Juravenator*. *Sciurumimus* is difficult to classify because it is clearly a juvenile animal. Yet most think it is related to megalosaurs and it clearly had feather-like filaments. In the first description of *Juravenator* it was believed to be featherless, but more detailed examination under UV light clearly shows that it had both feathers and scales. It was originally considered to be a close relative of *Compsognathus*; however, more recent studies suggest that its affinities lie with more advanced forms.

Compsognathus, even though it is known primarily from only a single specimen, is a seminal player in understanding non-avian dinosaurs. Ever since its discovery it has been studied by all the luminaries of dinosaur paleontology. We are lucky that it still exists. Most of the collection in the institution where it is housed, the Bayerische Staatssammlung in Munich, was destroyed during the Allied bombing of the city in the Second World War. Specimens that were lost include many important dinosaurs from Ernst Stromer's excavations in Africa, including the type specimen of *Spinosaurus*. The only reason that the *Compsognathus* specimen survived was that when it became apparent that the war was probably going to be lost, one of the curators took the specimen to his home in the countryside.

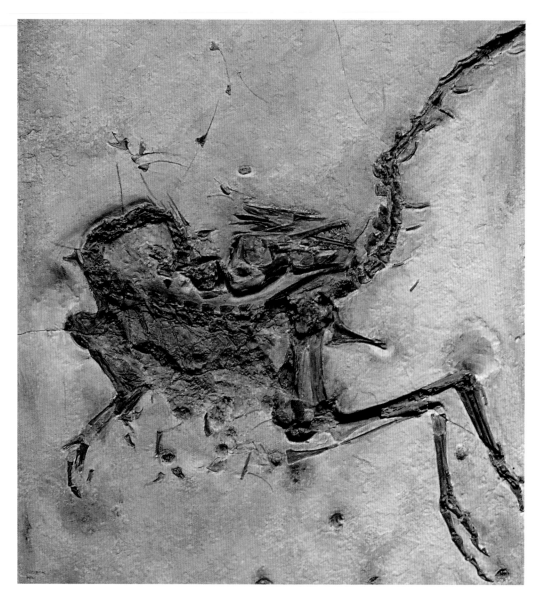

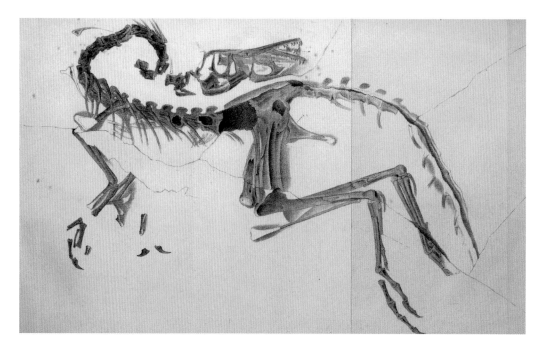

< The original specimen of *Compsognathus*. Unlike so many other specimens in the Munich museum, this slab survived the bombings of the Second World War. It was small enough to be taken to the countryside by one of the curators.

∟ An excellent lithograph of *Compsognathus*. The original stone currently resides at Yale University.

> This 1903 illustration of the stomach contents shows parts of the skull of a lizard ingested shortly before death.

∨ Some strange ideas have been floated about the use of the forelimbs in *Compsognathus*. One is that it used them as oars to swim and dive.

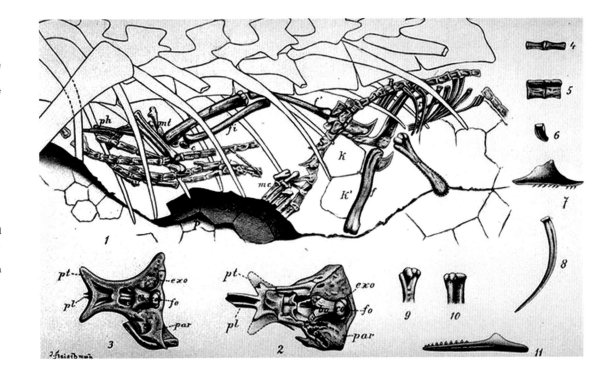

MONONYKUS OLECRANUS

LATE CRETACEOUS
NEMEGT FORMATION
MONGOLIA

IN 1990, DURING THE FIRST AMERICAN MUSEUM OF NATURAL HISTORY-MONGOLIAN EXPEDITION TO MONGOLIA, WE WERE IN THE CAPITAL OF ULAANBAATAR, BASED IN A DECREPIT BUILDING USED BY THE MONGOLIAN ACADEMY OF SCIENCES AS A STORAGE AREA FOR FOSSILS AND FIELD EQUIPMENT BEHIND THE NATIONAL NATURAL HISTORY MUSEUM.

Among Russian car parts, boxes of unlabelled fossils and various detritus, there were a couple of small offices. One of these belonged to Altangerel Perle, a Mongolian paleontologist who had been assigned to work with us by the Academy of Sciences. Among various fascinating specimens was one that was rather strange. We laid much of the small specimen out on the table; clearly it was something new and we were excited.

It had been collected by Mongolian paleontological expeditions a few years earlier at the Late Cretaceous Bugin Tsav locality. Bugin Tsav is a blisteringly hot, dusty, fly- and ant-infested Gobi Desert basin north-west of the Golden Mountain (Altan Ula) in Ömnögov Province, southern Mongolia. It is the site of many important dinosaur discoveries and this little dinosaur was one of them.

In front of us lay leg bones, hips, vertebrae, the back of a skull and importantly, forelimb bones. It was immediately clear from various bumps, proportions and processes of the limb bones that this was an animal very closely related to modern birds. Certainly it was an advanced coelurosaur. But what really took us by surprise were the forelimbs: instead of the typical long front limbs with three fingered hands, Perle's animal had a tiny, powerfully built, short limb that sported a single, extremely large claw.

A few months later Perle brought the specimen with him for an extended study trip to New York. Detailed analysis of it there demonstrated that it was, as initially thought, a coelurosaur. More analyses showed that it was related to a group of poorly preserved enigmatic dinosaurs called alvarezsaurs, which at the time were only known from the Cretaceous of Argentina; how these are related to other dinosaurs was at the time unknown.

We initially named the animal *Mononychus* in a paper in the journal *Nature*. Shortly thereafter we unfortunately found out that the name (meaning "single finger") had already been used for a beetle, so we had to change it to conform to the rules of scientific nomenclature, which (like thoroughbred race horses) require a unique name. We resolved this by simply exchanging the "ch" for a "k" to form *Mononykus*. In this paper, we proposed the bold hypothesis that *Mononykus* was intimately related to modern birds. At the time, this did not go down well, yet it is clear that *Mononykus* shares many similarities with modern birds not found in other non-avian dinosaurs. Most of these are subtleties of the boney skeleton, but a couple of important ones include the presence of a breastbone with a keel and the fusion of all of the wrist elements into a single block called the carpometacarpus. Our initial proposal has not stood the test of time and most now consider *Mononykus* to hold a more basal position, i.e. that it is not that closely related to birds, and several dinosaurs like troodontids and dromaeosaurs are much more closely related to the creatures that now populate our skies. But the story is still developing. Much is being learned about how *Mononykus* is related to other dinosaurs through fantastic discoveries of close *Mononykus* relatives.

∧ *Mononykus* portrayed as a
brightly coloured, feathered
creature.

And there is also the forelimb to consider. It is certainly one of the most unusual dinosaur forelimbs that has ever been recorded. In general, when we encounter really small, diminutive forelimbs (as in *Tyrannosaurus rex*), they are vestigial and almost useless, weak appendages. This is not the case for *Mononykus*. The forelimbs in this animal were very short, but stout and sturdy. The lower arm bone (the ulna) features a large olecranon process – basically the elbow. This indicates that significant leverage and force could be exerted by the forearm. This force would be transmitted through the hand via a wrist that was largely immobile and could move in only a couple of planes. This was topped off by the large claw on the hand, which must have been capable of a powerful digging force. Such highly modified appendages usually signify a very specialized lifestyle; animals with similar appendages today tend to be mammals like pangolins and anteaters. So, perhaps, the lifestyle of *Mononykus* and its relatives was not vastly different, and it used its powerful forelimbs to dismantle insect nests.

Joint AMNH–Mongolian expeditions in the mid-1990s recovered two new species of alvarezsaurs from the Ukhaa Tolgod beds. One of these is a large form called *Kol ghwa*, which is known from a single foot. It is about twice as large, and because of a paucity of fossils, little is known about it. We know much more about *Shuvuuia deserti*. In Mongolian, this means "bird of the desert", a reference to the birdlike nature of the skeleton. It is known from several specimens, both adults and juveniles, from several localities. Not only does it give us a better picture of what the closely related *Mononykus* looks like due to its well-preserved skull remains, it also provides other clues to the appearance of these animals. During preparation of the skull of the type specimen, numerous small, thin fibres were found, especially around the forelimbs and the skull. A few years later these filaments were subjected to a type of analysis that detects the presence of minute amounts of preserved protein; these analyses did indeed detect keratin, and specifically the type found in the feathers of modern birds. This provides powerful evidence that this advanced non-avian dinosaur, like other dinosaurs, was a feathered creature when alive.

∧ This skeleton of *Mononykus* is based on a composite of *Mononykus* and *Shuvuuia* bones.

> The closely related *Shuvuuia* is almost identical to *Mononykus*. It lived slightly earlier in time and occupied a desert habitat as opposed to a riverine floodplain.

∧ The powerful aberrant forearm of *Mononykus*.

OVIRAPTOR PHILOCERATOPS

LATE CRETACEOUS
DJADOKHTA FORMATION
CENTRAL ASIA

THE CENTRAL ASIATIC EXPEDITIONS WERE SOME OF THE MOST EXPENSIVE PALEONTOLOGICAL EXPEDITIONS EVER CONDUCTED. THEY WERE ORGANIZED BY THE AMERICAN MUSEUM OF NATURAL HISTORY AND OFFICIALLY TOOK PLACE FROM 1922 TO 1928. THESE EXPEDITIONS WERE THE BRAINCHILD OF HENRY FAIRFIELD OSBORN, THEN PRESIDENT OF THE MUSEUM, AND ROY CHAPMAN ANDREWS.

Roy Chapman Andrews was an intriguing figure. After graduation from Beloit College, Wisconsin, he began to pursue his dream of working at the Museum. He asked for a job and was given one tidying up around the mammalogy department and plying his self-taught taxidermy skills. Early in his career Andrews was recognized as both prodigal and a prodigy. Wanderlust and a natural ability to pick up foreign languages sent him out on expeditions; his first took him to the East Indies and to the Arctic. While pursuing an advanced degree he lived in a Japanese whaling village, taking some of the first quality moving pictures of marine mammals. During his time in the East, Andrews indulged himself in the food, the culture and all of the pleasures that Asia in the early twentieth century had to offer. Andrews would go on to become one of the greatest and best-known expeditioners and popularizers of science during this period.

He regularly appeared on radio shows, lectured frequently, penned over 20 books and wrote copious articles. Even though he is not considered a great scientist (he never finished his graduate education), Andrews was a charismatic frontman for what would arguably become the most famous paleontological expedition ever.

On his return from his initial sojourns to Asia, Andrews once again turned his eye towards the East. At the time the Museum was led by the powerful – albeit from today's perspective, controversial – leader, Henry Fairfield Osborn. Osborn founded the Museum's Department of Vertebrate Paleontology in 1891 and became President of the Museum Board of Trustees from 1908 to 1935. The scion of a wealthy and popular industrialist, Osborn had important connections in both society and commerce.

One of Osborn's many interests was the origin of humans. Although untested, as no significant fossils had been collected, a

> Although we don't have any direct fossil evidence for feathers in *Oviraptor*, their presence in close relatives strongly suggests their presence in this animal.

prevailing contemporary theory was that the origin of humans lay in Asia. However, few intellectuals in Europe or North America at the time would even consider that modern humans could possibly have originated on the Dark Continent. In his zeal for adventure, Andrews was able to convince Osborn to let him lead a series of expeditions to Central Asia, in search of the fossils that would corroborate Osborn's ideas.

The Central Asiatic Expeditions were grand undertakings and were also incredibly expensive, and here Osborn's connections and Andrews' charisma became key in persuading New York's wealthy to open their cheque books. The headquarters for these endeavours was an old royal palace near what is today Tiananmen Square in Beijing. These trips were innovative in the sense that instead of the typical transportation methods of camels and donkey carts, the Museum imported motorized vehicles that were supported by gasoline and parts sent out by camel caravan to makeshift supply depots deep in the Gobi Desert. These expeditions were truly modern in every way.

One of the most famous discoveries came at a locality called the Flaming Cliffs (Bayn Dzak); it was here that the first universally recognized dinosaur eggs were found, and this find is related to the discovery of the theropod dinosaur *Oviraptor* at Bayn Dzak.

During the 1923 instalment of the expedition, team member George Olson came upon an unusual fossil occurrence: the skeleton of a reasonably small, new kind of theropod dinosaur. While this was a fine discovery on its own, this dinosaur was found associated with several dinosaur eggs. Earlier in the season, the expedition had come across a number of eggs, sometimes even in nests. When the specimens arrived in New York, they generated huge excitement and international publicity (and even a newsreel). Shortly thereafter, in 1924, Henry Fairfield Osborn published a seminal paper on some of the Flaming Cliffs finds entitled "Three new Theropoda, *Protoceratops* zone, central Mongolia".

In addition to the iconic Mongolian *Velociraptor* (see p.106), Osborn named another dinosaur – the one found by Olson that was associated with the nests. Osborn named the animal *Oviraptor philoceratops*, which means "egg eater fond of ceratops". Why this unusual name? This is because when the Flaming Cliffs dinosaur eggs were first interpreted they were considered to be the eggs of the ornithischian herbivore *Protoceratops andrewsi* (see p.190). The chain of reasoning is that because dinosaur

< Roy Chapman Andrews, leader of the Central Asiatic Expeditions, with dinosaur eggs at Bayn Dzak in the early 1920s.

⌐ Henry Fairfield Osborn, founder of the Department of Vertebrate Paleontology and director of the Museum. It was his vision that sparked the Central Asiatic Expeditions.

∧ A camel resupply caravan arriving at Bayn Dzak (the Flaming Cliffs).

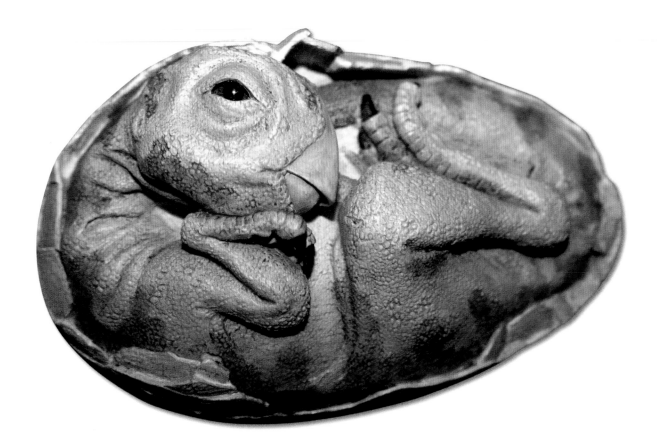

eggs of the type found with the *Oviraptor* are ubiquitous at the Flaming Cliffs, they must be the eggs of the most common dinosaur at the same locality. It was hence interpreted that the *Oviraptor* was predating on the *Protoceratops* nest when it met its demise. In all fairness, Osborn did qualify his analysis by stating he could not be absolutely certain that the eggs were those of *Protoceratops*. Nevertheless, this became scientific orthodoxy for decades.

This all changed in 1993 when the AMNH–Mongolian Academy of Sciences expeditions discovered a remarkable locality in southern Mongolia. Named Ukhaa Tolgod, the site has become one of the most important Late Cretaceous dinosaur localities in the world. The first fossils ever found at Ukhaa Tolgod were collected in July 1993. On that day I saw a dinosaur egg exposed on the ground a few paces in front of me. This was not an unusual occurrence because eggs are commonly found in these sediments. It was the type of egg that is most common at the Flaming Cliffs and the same as that found underneath the 1923 *Oviraptor philoceratops* skeleton. But when I picked it up I could clearly see it was special: it held an embryo of a developing dinosaur inside. Moreover, it was not a *Protoceratops*

embryo; it was a theropod dinosaur embryo on the half shell. At the time this was the first theropod dinosaur embryo known, and it conclusively demonstrated that the eggs at Flaming Cliffs were not laid by *Protoceratops*. Other discoveries at Ukhaa Tolgod of the closely related dinosaur *Citipati* add even more to the story (see p.98). The preparation and analysis of the embryo back in New York, showed that not only was it a theropod, it was an oviraptorosaur – the group of dinosaurs to which *Oviraptor* belongs.

Returning to the original *Oviraptor philoceratops*, recent analysis has told us even more about this unusual animal. Even though the type specimen has been studied intensively for nearly 100 years, something was missed. Intermixed with the adult skeleton and the eggshell fragments and partial eggs, are the bones of tiny oviraptorosaurs. Although we cannot demonstrate it definitively, the *Oviraptor* was probably a parent tending to its nest of hatched chicks when they perished. In this highly supported interpretation, instead of a predator, it was a good parent. Unfortunately, the strict rules of scientific nomenclature do not allow us to rename it, and "egg eater" it remains.

∧ A sculptural representation of an oviraptorid embryo in its egg. In 1993 fossils were found showing *Oviraptor* was not an egg thief.

⊐ A drawing of the original *Oviraptor* fossil. Interestingly, the numerous egg shell pieces found associated with the specimen are not illustrated.

> *Oviraptor* had fairly long, curved claws on its feet, but lacked the hooked raptorial claws of *Velociraptor* and its relatives (compare with the claw on p.106).

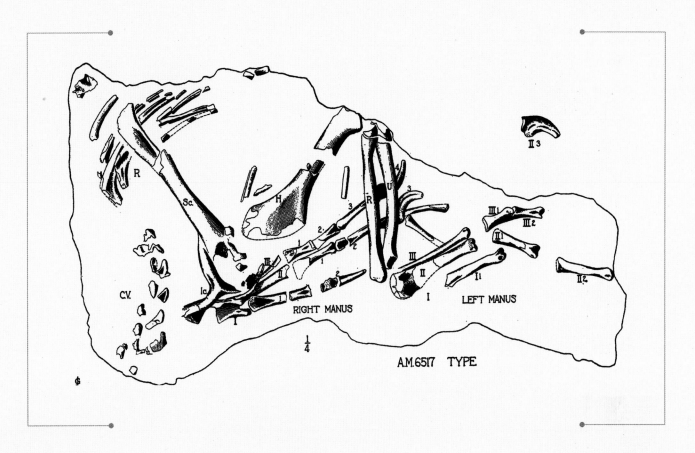

KHAAN MCKENNAI

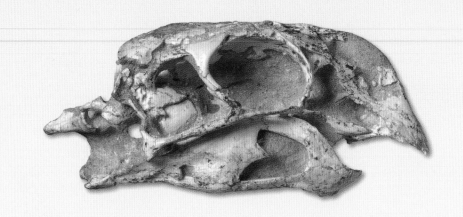

LATE CRETACEOUS
DJADOKTHA FORMATION
CENTRAL ASIA (MONGOLIA)

ONCE CONSIDERED TO BE THE RAREST OF DINOSAURS, IT IS NOW APPARENT THAT OVIRAPTOROSAURS WERE DIVERSE AND ABUNDANT THROUGHOUT THE CRETACEOUS (ESPECIALLY THE LATE CRETACEOUS) IN THE NORTHERN HEMISPHERE. CENTRAL ASIA IS THE CENTRE OF DIVERSITY WITH SEVERAL SPECIES, INCLUDING *CITIPATI* (SEE P.98) AND *OVIRAPTOR* (SEE P.90).

*K*haan was the first oviraptorosaur to be collected at the Ukhaa Tolgod locality in Mongolia by Museum and Mongolian paleontologists. Unlike the other two oviraptorosaur specimens described in this book, *Khaan* lacked a large cassowary-like crest on its head. It was also fairly small at under 2m (6½ft) in length.

There are multiple signs that *Khaan* was a social animal. The fossils at the Ukhaa Tolgod locality were killed by the event that preserved them, with evidence pointing to the collapse of large, water-laden sand dunes that liquefied and buried animals alive. In July 1995, the tail of a *Khaan* specimen (the first one had been found two years earlier at the same locality) was found protruding from a small hill. Usually, in this sort of case, when you dig into the hill you are likely to find the rest of the tail. On this occasion however, a beautifully preserved specimen of this small dinosaur was collected.

During the excavation another small dinosaur – also a *Khaan* specimen – was found lying next to it. Both animals had been killed by the same event and exhibited clear signs of

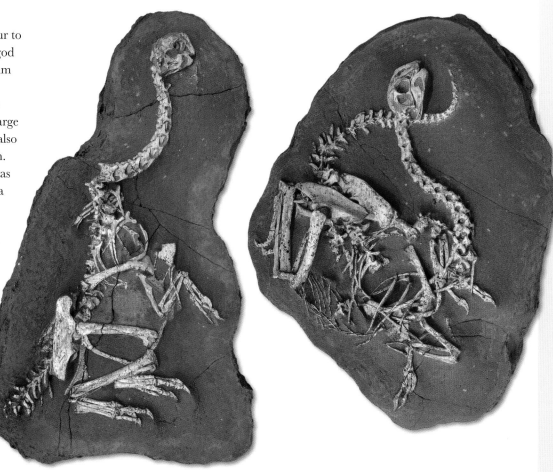

∧ The spectacular *Khaan* specimens "Sid and Nancy" from Ukhaa Tolgod.

broken necks. The specimens were nicknamed "Sid and Nancy" after the bassist of the Sex Pistols and his girlfriend Nancy Spungen, who met her demise at Manhattan's Chelsea Hotel in 1978. Two decades later, the specimen was included in a larger study on determining the sex of theropod dinosaurs in general. What had been determined was that the shape of the bottom process of an anterior tail vertebra differed in male and female crocodiles. The same conditions were found in Sid and Nancy, strongly indicating that they were a male and female pair when alive.

Large associations of immature *Khaan* specimens have also been found at Ukhaa Tolgod. What is interesting about these

associations is that, like Sid and Nancy, they are believed to have perished in a single event. These specimens are all about the same size, showing that they were an age cohort that was living in a flock. This intensely social behaviour may be a characteristic of the entire group.

Other oviraptorosaurs have also been found in bonebeds, including a peculiar example from the Mongolian locality of Khulson. There are dozens of specimens of juveniles less than a year old, combined with at least three fully grown individuals. This suggests that like living ostriches, these were very social animals that did not just brood their nests but cared for their young, and enjoyed a long family life together.

∧ A *Khaan* bonebed from Ukhaa Tolgod. Such bonebeds, this one containing at least seven individuals, suggest that these animals were social in life.

⌐ The type specimen of *Khaan*, from the Late Cretaceous Ukhaa Tolgod beds of Mongolia.

CITIPATI OSMOLSKAE

LATE CRETACEOUS
DJADOKHTA FORMATION
MONGOLIA

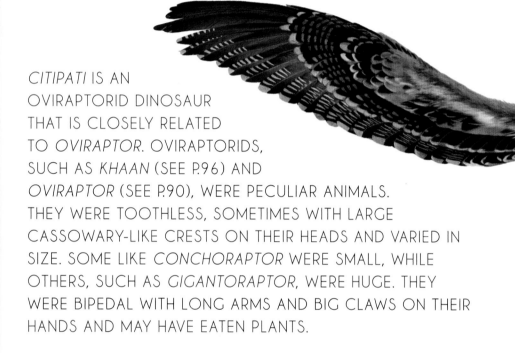

CITIPATI IS AN
OVIRAPTORID DINOSAUR
THAT IS CLOSELY RELATED
TO *OVIRAPTOR*. OVIRAPTORIDS,
SUCH AS *KHAAN* (SEE P.96) AND
OVIRAPTOR (SEE P.90), WERE PECULIAR ANIMALS.
THEY WERE TOOTHLESS, SOMETIMES WITH LARGE
CASSOWARY-LIKE CRESTS ON THEIR HEADS AND VARIED IN
SIZE. SOME LIKE *CONCHORAPTOR* WERE SMALL, WHILE
OTHERS, SUCH AS *GIGANTORAPTOR*, WERE HUGE. THEY
WERE BIPEDAL WITH LONG ARMS AND BIG CLAWS ON THEIR
HANDS AND MAY HAVE EATEN PLANTS.

So far *Citipati* has only been found in the bright red Late Cretaceous (about 78 million years ago) sandstones of Ukhaa Tolgod, Mongolia. The first specimen was collected in 1994 and several other individuals have since been found. The somewhat unusual name comes from the Himalayan Buddhist deities that are the guardians of the funeral pyre. Fittingly the Citipati are usually portrayed as a pair of dancing skeletons surrounded by a halo of flames. The specific name is in honour of Halszka Osmólska, one of the leaders of the Polish–Mongolian Expeditions of the

1960s and 70s, which excavated so many important Gobi Desert dinosaurs.

Specimens of oviraptorid dinosaurs are very common at Ukhaa Tolgod. At least two species have been found there, both of which were new to science when we discovered the locality. Besides *Citipati*, we have excavated several other specimens of a smaller but closely related form. This species, called *Khaan mckennai*, is known from numerous individuals.

A characteristic of oviraptorids is that they appear to be gregarious; many specimens collected at Ukhaa Tolgod and

the circumstances of their preservation supports this. One of the reasons that specimens from the Ukhaa Tolgod locality are so well-preserved is that they were apparently buried alive and have been protected by the very agent that killed them. How did this happen?

This high-quality preservation of fossils is seen at many localities in the Djadokhta Formation of Mongolia and has been debated for decades. The usual explanation employed is that these animals were overcome by powerful sandstorms, but there are problems with this scenario. One is

∧ A fully feathered pacing adult *Citipati*, the most common theropod at the Ukhaa Tolgod locality.

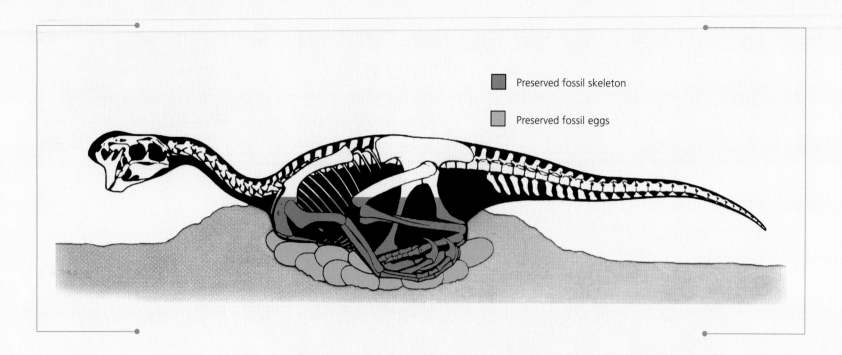

Preserved fossil skeleton

Preserved fossil eggs

that sandstorms strong enough to completely overpower and bury human-sized animals, let alone larger dinosaurs, have never been observed in the modern world – even in the harshest and windiest of places.

Our geological teams working at Ukhaa Tolgod reanalyzed the subject. By carefully mapping out the exact places in which vertebrate fossils were found, they observed two types of what geologists call facies, or rock strata with particular characteristics of appearance, composition or conditions of formation. In one of these facies there are no fossils, but there are tell-tale signs that these rocks were formed from large ancient sand dunes. A sand dune signature is composed of steep bedding planes known as cross-beds that show the shifting surfaces of wind-blown sand dunes. The second type of facies contains a great many fossils but no cross-beds. In addition to this, one occasionally finds fist-sized cobbles that would have been far too heavy to have been moved by wind. This second facies is described as "structureless" because of the lack of cross-beds. So why do the cross-bedded rocks contain no fossils while the structureless sands do? Detailed analysis of the individual sand grains provides a clue: these sands contain a high clay component, and clay absorbs water.

The geologist's interpretation of the environment is thus a relatively open, arid area of sand dunes with some interdune areas.

Occasional ephemeral ponds of water were also present. In rare occurrences, perhaps over hundreds of years, very large storms would occur. Because of the clay content in the dune sands, the water was absorbed instead of draining quickly as is true of most sandy soils. Compounding this was the presence of plants that stabilized the dunes to some extent; we know about the existence of plants because of the tell-tale root marks, called rhizoliths, left in the sediments. As the sand dunes absorbed more and more water, they eventually went past a critical point, becoming unstable and collapsing into a sandy, liquid river. Something similar happens when you are at the beach and the sand in your sand castle becomes just a little too wet, and rivers of sand pour over the walls. This sand river was what entombed the animals at Ukhaa Tolgod.

This remarkable preservation, where animals are buried alive, accounts for one of the most spectacular fossils I have ever found. During the 1993 instalment of the American Museum of Natural History–Mongolian Academy of Sciences expedition, I found the claws of a hand protruding from the bright red sandstone. I marked it and asked some members of the team to investigate it while I went out to take care of another excavation. A while later, a car approached my location. An excited team member jumped out and said, "There are eggs under the skeleton." This specimen would eventually become known as "Big Mama",

∧ A schematic interpretation of the "Big Mama" nesting *Citipati* specimen opposite.

> The nesting *Citipati* specimen named "Big Mama" in plan view. This offered clear indication that non-avian dinosaurs brooded their nests.

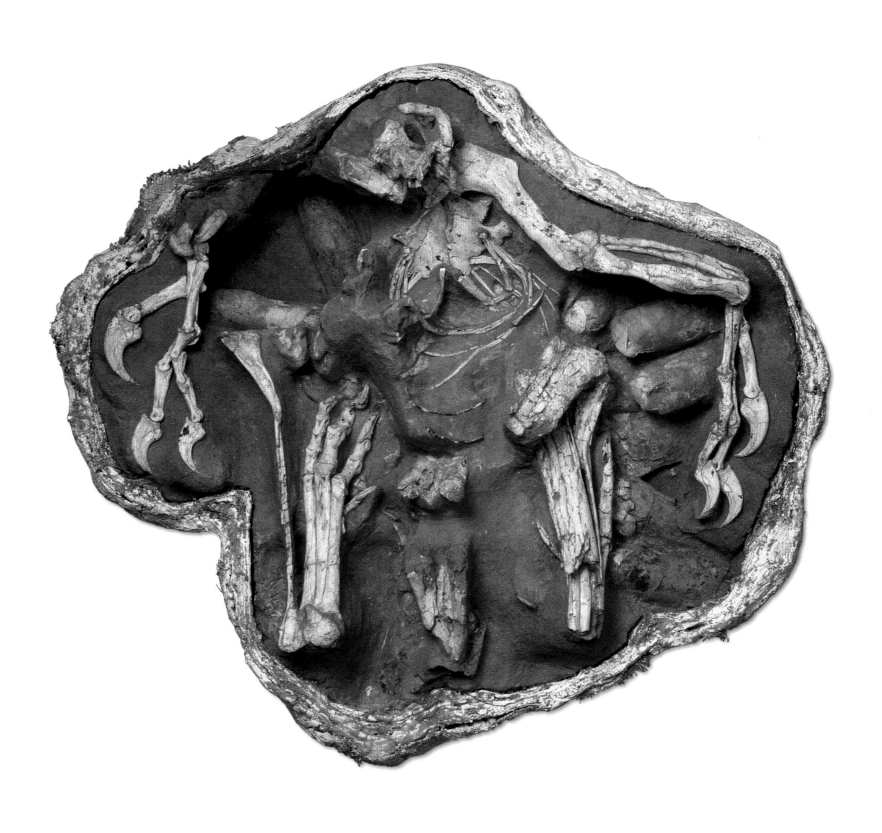

∧ The Ukhaa Tolgod locality. This area is called the Camel's Humps amphitheatre. It is where many important dinosaurs have been discovered.

< The *Citipati* embryo from Ukhaa Tolgod. This specimen confirmed that the eggs associated with the *Oviraptor* specimen were not *Protoceratops* and that these animals were parents, not egg-eaters.

and is an adult *Citipati* sitting on a nest of eggs. This generated worldwide interest because it provided smoking-gun evidence that non-avian dinosaurs sat on their nests and brooded their eggs just like modern birds.

This discovery, together with two further finds at the same locality, raises some interesting questions and points to some profound conclusions. An obvious one is that the typical behaviour of nest-brooding that we associate with modern birds has its roots fairly deep in non-avian dinosaur history. But there are two peculiar things about this. The first is why *Citipati* and its very close relatives are the only dinosaurs for which we have direct fossil evidence showing this behaviour. This is certainly a conundrum, but it probably relates to the fact that oviraptorosaurs happen to be the most common theropod dinosaurs at the localities where these sand flows have entombed live dinosaurs. Judging from the high frequency of these dinosaur eggs and their fragments at both Ukhaa Tolgod and Bayn Dzak (where the *Oviraptor* was found in the 1920s by Museum paleontologists), these were very common animals which nested in the microhabitats that they frequented.

ORNITHOLESTES HERMANNI

JURASSIC
MORRISON FORMATION
WESTERN NORTH AMERICA

AS THE NAME IMPLIES (IT MEANS "BIRD ROBBER"), *ORNITHOLESTES* WAS RECOGNIZED AS HAVING BIRDLIKE FEATURES SOON AFTER ITS DISCOVERY IN EASTERN WYOMING, USA, IN 1900. IT WAS A SMALL, AGILE ANIMAL, ABOUT 2M (6½FT) LONG. INDEED, THE ANATOMY OF THIS ANIMAL WAS SO BIRDLIKE IT EVEN PROMPTED ENGLISH ORNITHOLOGIST PERCY LOWE TO SUGGEST, IN 1944, THAT IT WAS FEATHERED WHILE ALIVE.

Ignored for decades, this prescient statement is now supported by phylogenetic evidence. It was discovered by American Museum of Natural History expeditions at a place they named Bone Cabin Quarry, because a shepherd had built a small dwelling almost completely out of dinosaur bones at the site. The location is near where the famous discoveries at Como Bluff had been made by O.C.

Marsh and E.D. Cope a few decades earlier. It is still known only from one unique specimen. Another specimen in the Museum collections, a partial hand, had been attributed to this dinosaur, but new research shows that it probably belongs to a very similar coeval theropod called *Tanycolagreus*. This was one of the first small animals to be described in what would later be recognized as the advanced dinosaur

subgroup Coelurosauria, which also includes living birds.

In early reconstructions *Ornitholestes* is often depicted with a small horn at the end of its skull. In fact, this is not the case, as during the fossilization process the skull was deformed, and bones on the right side slid over the bones on the left when the skull was flattened. When animals fossilize, they are almost never perfectly preserved. Let's examine this further: after an animal dies, it is subject to all sorts of depredations that make fossilization an extremely unlikely event. For fossilization to occur, an animal needs to be buried quickly before it is consumed by scavengers, or its skeleton is destroyed by environmental influences. Certain environments preserve bodies better than others. Arid areas are very good, much better than tropical humid ones, where animals decay extremely quickly, unless they are buried by sudden events, such as volcanic eruptions or flash floods. Big animals, because their skeletons are more durable, are also much more likely to be preserved than smaller ones. At Bone Cabin Quarry *Ornitholestes* is the only relatively

∧ A fully feathered illustration of *Ornitholestes*. It is a primitive representative of coelurosaurs, the group that also includes birds. Many bird traits in the skeleton are already present.

∟ The skull of *Ornitholestes*. The "horn" at the end is the result of deformation of the skull during fossilization.

↘ How *Ornitholestes* is commonly portrayed. Today the fossil mount would look much more like the *Deinonychus* reconstructed on page 111.

well-preserved specimen of a small animal that was collected.

These biases give us a highly flawed and skewed impression of the quality of the fossil record. Specifically, we know very little about small animals, and especially those that live in the most speciose terrestrial environments on Earth – tropical rainforests. There are probably at least three orders of magnitude more species living in the Amazon Basin today than live in the Great Plains of North America. Sometimes, however, we get a little lucky. In the last few years, Early Cretaceous fossils have been discovered in northern Myanmar, near the Chinese border. In some of these sites pieces of amber, the remains of all sorts of insects, other arthropods, and plants are common. Very occasionally these preserve amphibians and reptiles – frogs and

lizards – that look astoundingly modern in appearance. On even rarer occasions the fossils of feathers and partial bodies of basal avians and dinosaurs have been reported. This is an incredibly important locality that is just beginning to be studied. It provides us with a unique picture of a diverse tropical habitat during the Early Cretaceous.

In comparison, *Ornitholestes'* environment was relatively depauperate. Although we do not have a lot of evidence, what we do have in the form of relatively fragmentary pieces of bones and teeth is that the environment in which *Ornitholestes* hunted was a well-drained floodplain, punctuated by sparse forests. The rivers were full of crocodiles that would look very modern in appearance, as well

as lungfish and turtles. Primitive birds and pterosaurs flew in the sky, extraordinarily large sauropod dinosaurs fed on shrubs and forest plants, and stegosaurs ambled around. Even lizards and primitive mammals abounded. While not nearly as speciose as tropical areas today, it would have been an amazing vista to behold.

VELOCIRAPTOR MONGOLIENSIS

ASIDE FROM *TYRANNOSAURUS REX*, *VELOCIRAPTOR* IS ONE OF THE WORLD'S BEST-KNOWN DINOSAURS. OF COURSE, THIS HAS EVERYTHING TO DO WITH ITS STARRING ROLE IN SEVERAL INSTALMENTS OF THE *JURASSIC PARK* MOVIE FRANCHISE.

It was probably cast because one of the defining characters of deinonychosaurs (the subgroup to which it belongs) is so spectacular – the large raptorial claw on the second toe of each foot. Before getting into specifics about this remarkable animal, it is useful to contrast the film image of *Velociraptor* in the early films with the way we think of it now. If you can recall the look and behaviour of the first image of the *Jurassic Park Velociraptor*, it was a very reptilian, human-sized, group-hunting stalker in a gloomy, poorly lit laboratory.

What is wrong with this picture? First, the size of the animal was incorrect: it was not the size of a human. If you had been in that *Jurassic Park* laboratory and turned that corner, it is likely the *Velociraptor* pack would have fled. An adult *Velociraptor* was actually about the length of a medium-sized coyote, and most of that was tail. The *Velociraptor* skull was the size of a very large fox and perhaps only a threat to a rabbit. So contrary to *Jurassic Park*, *Velociraptor* would be about as dangerous to humans as an annoying dachshund, and easily neutralized.

The first *Velociraptor* specimen was discovered by the American Museum of Natural History Central Asiatic Expeditions at the famous Bayn Dzak or Flaming Cliffs locality in central Mongolia in August

1923. It was described in a very short paper by the then President of the Museum, Henry Fairfield Osborn, who was the chief motivator behind the expeditions. In his description he presciently observed that this is a "bird-like dinosaur". He did not know how right he was, as 70 or so years later *Velociraptor mongoliensis* would be one of the animals to hold centre stage in the burgeoning, sometimes contentious, debate on the origin of birds.

Since the original discovery of *Velociraptor*, other spectacular specimens have been discovered – all from Mongolia's Gobi Desert. Some of the new finds have told us much about its lifestyle. The ultimate specimen has to be the so-called "fighting

dinosaurs" from the locality of Tugrugeen Shireeh in Central Mongolia, adjacent to the Bayn Dzak locality. Found by a joint expedition of Mongolian and Polish paleontologists in 1971, it is a moment frozen in time from about 80 million years ago.

∧ The raptorial second digit of the dromaeosaur *Velociraptor*. In life, the claw would have been about twice as big, because a keratinous claw sheath covered the bony core.

> The fabulous "fighting dinosaurs" specimen, one of the great treasures of paleontology.

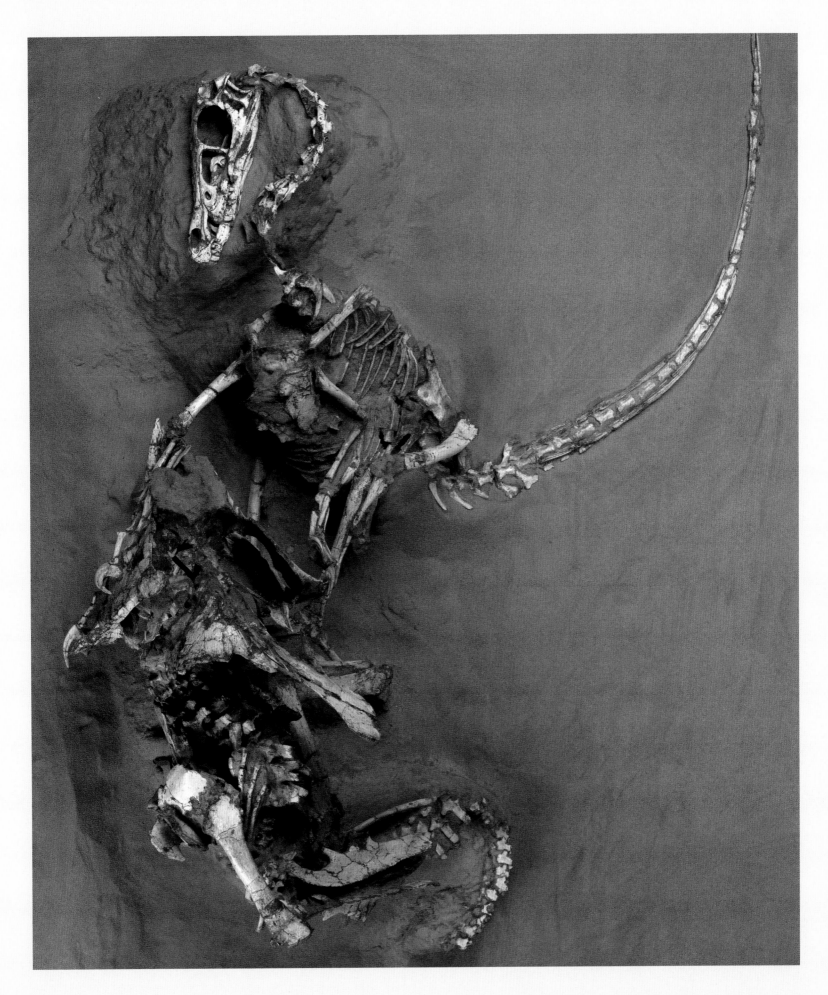

The specimen consists of a *Velociraptor mongoliensis* entangled with a *Protoceratops andrewsi* specimen. *Protoceratops* (see p.190) is an herbivorous dinosaur, and as an adult would have been about the size of a large pig. It was probably the ecological equivalent of the sheep that are predated on by wolves in Mongolia's rural ecosystem today. Because of the nature of preservation in these beds (see *Citipati*, p.100), there is strong evidence that they were buried alive. What is preserved in this instance is an adult *Velociraptor* seemingly in combat with a *Protoceratops*. The large raptorial claw is embedded in an area that would have covered the important blood vessels feeding the head of *Protoceratops*. *Velociraptor*'s right arm is in its mouth, and the hand with its sharp talons is tearing the face while the forearm is being crushed. Without doubt this is the smoking gun of a predation event that happened about 80 million years ago.

Velociraptor has several characteristics that provide evidence of its close affinity to birds. It has a wishbone (see p.227), large hollow air sinuses in its skull, a swivel wrist, an S-shaped neck, and three primary toes on the foot that all face forward. We also consider *Velociraptor* to be feathered, and there are two strands of evidence to support this. The first uses the phylogenetic method. There is firm evidence that both modern birds, as well as some very close *Velociraptor* relatives (some of them featured in this book), not had only feathers,

but feathers of modern aspect – just like living birds. Since all of these animals for which we have direct evidence of feathers are descended from a single common ancestor, the presence of feathers is best explained as being present in their common ancestor. Since *Velociraptor* is part of this group, we would predict that it would have feathers in lieu of other evidence. But we do have further evidence that validates this prediction.

In 1994, our team collected a good *Velociraptor* specimen at Khulsan in the Gobi Desert. After preparation, it was noticed that on the ulna (the bone on the underside of the forearm) there is a row of small bumps. The ulna makes up the majority of the wing in modern birds and supports most of the primary flight feathers of the airfoil. If you ever see birds in flight, especially soaring birds like vultures or albatrosses, you know that the flight feathers are capable of making micro-adjustments to optimize performance. This is similar to the way in which sailors trim sails in the wind to optimize the speed of a boat. In living birds these bumps are not just attachment points for the feathers themselves, but are also pivot points for feather movement, and we found the same structures on the non-flying *Velociraptor* specimen! Since then these small bumps have been found in a variety of different theropod dinosaurs, including some larger ones such as *Concavenator*, a 6m (19½ft) long animal from Spain. What was the function

of the bumps since these animals could not fly? Many living birds, such as ostriches and game birds, regularly exercise their feathers, sometimes as elaborate territorial bluffery or as a dance to attract mates.

Finally, we have some evidence for what *Velociraptor* ate – at least some of the time. It is generally very difficult to determine paleo-diets. The most accurate way is to find a fossil animal with its last meal preserved inside its body cavity. It is a pretty sure bet that *Velociraptor* was a carnivore; its sharp teeth and claws are adaptations for this. However, in one fortuitous case – a specimen that was collected just a few years ago by a Japanese expedition – the *Velociraptor's* last meal was determined to be a pterosaur, a winged reptile that had not been able to flap away fast enough.

∧ The type specimen of *Velociraptor mongoliensis*. It was collected in 1923 and its significance was immediately recognized.

< A dromaeosaur rendered in a life-like pose. These animals were smart, active and feathered. Quite the departure from early lizard-like renderings.

∨ The radius (one of the forearm bones) of *Velociraptor*. This specimen shows small quill knobs. In modern birds, similar quill knobs are pivot points for the arm (wing) feathers.

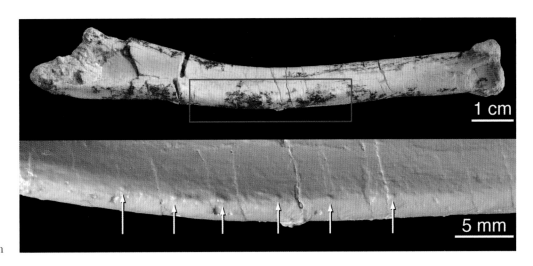

DEINONYCHUS ANTIRRHOPUS

EARLY CRETACEOUS
CLOVERLY FORMATION
WESTERN NORTH AMERICA

PERHAPS *DEINONYCHUS* SHOULD BE AS WELL (OR EVEN BETTER KNOWN) AS *VELOCIRAPTOR*. WHILE THE NAME MAY NOT BE AS CATCHY, IT IS, AFTER ALL, WHAT THE "RAPTORS" OF *JURASSIC PARK* WERE MODELLED ON – EVEN THOUGH IN THE MOVIE THEY WERE CALLED *VELOCIRAPTOR*.

*D*einonychus is a fascinating animal with several good backstories, and it provides us with an introduction to one of the most fascinating groups of theropod dinosaurs, one that is inextricably linked, for both historic and anatomical reasons, to the origin of modern birds.

Deinonychus was named in 1969 by Yale paleontologist John Ostrom, who had found remains of the animal in 1964 near Billings, Montana. Ostrom had been a student at the American Museum of Natural History in the 1950s where he came in contact with Barnum Brown, who was retired but still in residence at the Museum. In the 1930s Brown collected the remains of several small carnivorous dinosaurs from the Cloverly Formation, where Ostrom would later find his specimens. The specimen Brown collected was a fairly complete skeleton, and he set to work on what he informally named "*Daptosaurus*", and had intricate drawings produced for it for future publication.

Whether due to the Great Depression or the Second World War, Brown never formally published his work. All that remains of the link between Brown and Ostrom is a

faded letter in the archives of the Museum in Brown's handwriting: scribbled across the top is an annotation that says "for Ostrom". In it are the directions to the site where Brown found the "*Daptosaurus*" and Ostrom would later "discover" *Deinonychus*. Sadly, Brown's contribution was never recognized in any of the papers produced by Ostrom on this remarkable animal.

Like almost all other dromaeosaurs, *Deinonychus* was fairly small (about 2.5m/8¼ft) and bipedal. It also displays a slew of distinctive features. Probably the most obvious is the presence of a very large claw on the second digit of the foot. As alluded to earlier (see *Velociraptor*, p.106), this large claw was undoubtedly used for dispatching prey. These were large-brained, smart, cursorial animals – active predators in their time. But there are subtleties in their skeletons that show something very special, namely boney characteristics that align these animals closely with modern birds.

While the idea that birds are a type of dinosaur goes back to the studies of Thomas Huxley (see p.20), John Ostrom championed this idea which would gain

nearly universal acceptance in the years to come. One of the more important characteristics that Ostrom examined was the shape of bones in the wrist of the animal. Previously he had done extensive work on the early avialan *Archaeopteryx*. During these studies he noticed that there was a crescent-shaped bone in the wrist that allowed the hand of the animal to fold back when it was alive. You can see this motion in a pigeon: when it extends its arm (wing) it folds out like an accordion, extending the hand at the same time. The pivot point for this movement is the smooth, crescent-shaped wrist bone, which allows the hand to only move in one plane. This is a crucial adaptation that set the stage for the evolution of a rigid wing.

The crescent-shaped wrist is not the only boney character that is shared by *Archaeopteryx*, dromaeosaurs like *Deinonychus* and modern birds. These are discussed in detail elsewhere, but here it will suffice to point out that dromaeosaurs and many other advanced theropods also have wishbones, boney

breastbones, hollow bones, and feet with three toes that all point forwards.

Ostrom's work on *Deinonychus* was one of the key components of what is called "the dinosaur renaissance". Most of the work on dinosaurs up to this point both thought of and illustrated these animals as sluggish, lumbering reptiles. In addition, the number of paleontologists working on reptiles paled in comparison to the number working on mammals and fishes, but this all changed and changed quickly. The shift is perhaps best exemplified by an illustration of *Deinonychus* created by Ostrom's student Robert Bakker, which graced the cover of *Scientific American* in April 1975: instead of a tail-dragging "reptile", the illustration featured a highly active, running dromaeosaur perched on one leg in pursuit of its prey.

Work began on reimaging dinosaurs in a biologically constrained scientific framework. Much progress was made and ideas about warm-blooded dinosaurs, parental care and active, intelligent animals were formulated. But no one was prepared for what the end

of the millennium would bring in terms of discovery; it was at this time that the great fossil fields of Northeastern China began to produce extraordinary fossils which would provide concrete proof for many of these new ideas and would change the image of dinosaurs forever.

In 1996 at the Society of Vertebrate Paleontology meetings in New York City, Chen Pei Ji, an invertebrate paleontologist from Nanjing, brought some photographs of a new small dinosaur specimen (about 75cm/29½in) from Liaoning province. It came from rocks that were known to be about 130 million years old. Although blurry, the photos

∧ A mount of *Deinonychus* in the Museum. Note the long tail that is stiffened by forward and backward projections of each vertebra.

< The formidable right foot of *Deinonychus*.

∧ This early reconstruction of the skull of *Deinonychus* is based on the more primitive theropod *Allosaurus*. We now know that the skull was much more lightly built and that the animal had binocular vision.

∨ This 1969 drawing of *Deinonychus* has become an iconic image. Showing this animal in full stride, it illustrates an active and engaged predator. It has become a moniker for what has been called "the dinosaur renaissance".

clearly showed the presence of filament-like fibres on the animal. These would prove to be primitive feathers – the first concrete evidence of a non-avian dinosaur with feathers. Named *Sinosauropteryx*, this little animal was only a footnote in what was to come.

Sinosauropteryx is a compsognathid dinosaur, meaning that it is a member of a group that is related to birds, but not as closely related as dromaeosaurs. Initial inspection of the specimen showed the filaments, which resembled a Mohawk haircut, over the whole body. Yet this is only an illusion, as the specimen preserved on a flat rock surface is the trace of a three-dimensional animal in two dimensions. However, subsequent study of the animal clearly showed that the filaments were distributed across the entire surface of the animal. Several argued that these were primitive feathers, many vehemently disagreed.

New evidence came at a fast and furious pace. The first was the discovery of *Caudipteryx*, and *Protarchaeopteryx*. Both of these remarkable specimens were covered in feathers of modern aspect, which have a central supporting spine called a rachis and individual filaments emanating from it that form the vanes, like those seen on a modern bird. Unlike *Sinosauropteryx*, everyone agreed that these were feathered animals. However, those who disagreed with the "birds are dinosaurs" hypothesis argued that these animals were not dinosaurs, but flightless birds. While they were not dromaeosaurs, at least *Caudipteryx*, which has a complete covering of feathers including large feathers on the arm and a tail fan, can be definitively assigned to the oviraptor group. Further they argued the filaments associated with *Sinosauropteryx* were the by-products of the decomposition of muscles and tendons.

But a few new specimens definitively showed that one of these is a dromaeosaur. The specimen is nicknamed "Dave". It is a very well preserved dromaeosaur, referable to the genus *Sinornithosaurus*. The specimen, in the collections of the National Geological Museum of China, is preserved on two slabs (see p.225). It is what we call "bookmarked": imagine if a pigeon was shut in a very heavy book, compressed and left for years. Then, when the book is opened to the right pages, there would be an impression of feathers and bones on each facing page. In

∧ A reconstruction of the *Sinornithosaurus* specimen called "Dave". Direct fossil evidence shows that it was completely feathered with some feathers (on the arm) just like those in modern birds.

the case of Dave, what emerged from the finely laminated stone were the inside halves of a remarkably preserved dinosaur entombed in the style of a crucifixion, with its entire body covered in feathers. Some of these are primitive, single fibres, while others, especially those on the arms, are feathers of modern aspect.

Since the discovery of Dave, literally hundreds, if not thousands, of dromaeosaur

and troodontid fossils have been recovered from rocks that range from Middle Jurassic to Early Cretaceous in Northeastern China. These include some fantastic animals that are completely covered in feathers. The presence of feathers was predicted a long time ago, but no one was prepared for how the discovery of these fossils would change our perception of these animals.

SAURORNITHOIDES MONGOLIENSIS

LATE CRETACEOUS
DJADOKHTA FORMATION
MONGOLIA

THE AMERICAN MUSEUM OF NATURAL HISTORY'S CENTRAL ASIATIC EXPEDITION WAS THE FIRST HIGHLY ORGANIZED PALEONTOLOGICAL EXPEDITION TO MONGOLIA'S GOBI DESERT. THESE EXPEDITIONS (WHICH RAN BETWEEN 1922 AND 1928) WERE SOME OF THE LARGEST SCIENTIFIC UNDERTAKINGS OF THEIR TIME.

A great deal of money was spent, and everything, from automobiles to field supplies, was sent from the USA to supply the expedition. Based in an old royal palace near what is today Tiananmen Square in Beijing, Museum scientists drove across the northern plains of China towards remote areas and into Mongolia itself. Previously camel caravans carrying car parts, gasoline and other equipment had created supply bases along the caravan routes. Their journey took them to Urga (now Ulaanbaatar), the capital of Mongolia, where the necessary permits were secured, and from there they travelled to the fossil fields.

The motive for this massive undertaking was to pursue a pet theory of Museum president Henry Fairfield Osborn that humans originated in Asia. While the Central Asiatic Expeditions did not succeed in finding fossils relevant to the origins of our own species, they did uncover some of the most important dinosaur specimens to date and opened up a whole new area to paleontological exploration. Together with *Velociraptor*, the first well-documented dinosaur eggs and nests, and *Oviraptor*, one of their greatest discoveries was the small theropod dinosaur *Saurornithoides*.

At the time there was a great deal of confusion about troodontids. For dinosaurs that belong to the primarily carnivorous Theropoda, the teeth are unusual. Most species have teeth with large denticles, which are usually associated with plant-eating animals. The first troodontid remains were collected in the American West in the nineteenth century based on such teeth. Because the teeth were atypical of carnivorous dinosaurs, they were confused with the herbivorous, contemporaneous taxon *Stegoceras*. *Stegoceras* is an ornithischian dinosaur, a member of the dome-headed pachycephalosaurian group. It was not until the early 1900s that this confusion was resolved. Clearly this was recognized in the naming of *Saurornithoides*; as the name implies – it means "reptile bird" – this is a very bird-like, non-avian dinosaur.

Troodontids do not appear to have been very diverse. By far the most common troodontid is a small animal called *Anchiornis* from the Late Jurassic of Northeastern China. Hundreds of these specimens have been found and they display a number of birdlike characteristics. The primary example is the fossil presence of a complete set of feathers. These

∧ Dinosaur science leads to fanciful claims. This is a recreation of what might have happened if dinosaurs with big brains, like Troodon, had survived the asteroid impact. Perhaps it would have evolved into a large brained biped?

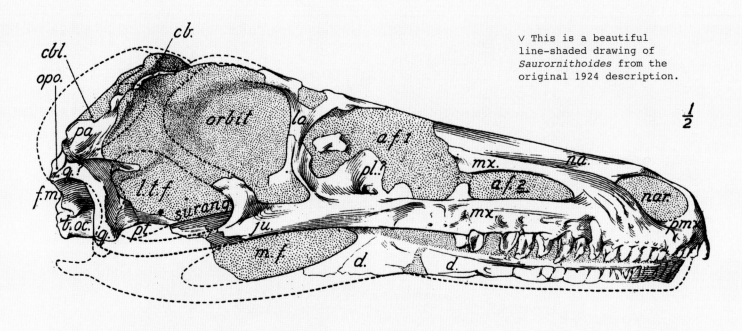

∨ This is a beautiful
line-shaded drawing of
Saurornithoides from the
original 1924 description.

½

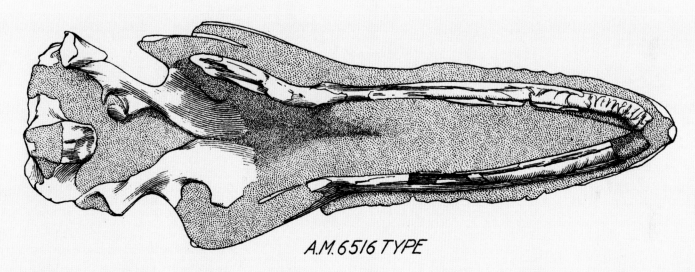

A.M. 6516 TYPE

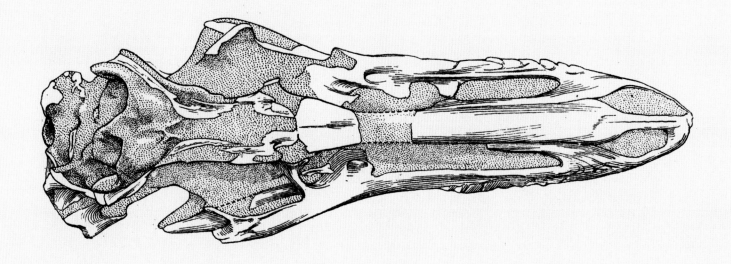

feathers are both plumulaceous (downy, those most associated with a thermal barrier) and pennaceous (those with a stiff shaft down the axis of the feather); pennaceous feathers are usually associated with display and aerodynamics. What makes the *Anchiornis* feathers so interesting is not necessarily their structure, but their distribution. It was surprising enough when a feathered, wing-like foil was found on the forelimb, but even more so when a reciprocal structure was found on the hind limb. As in the dromaeosaur *Microraptor* (see p.125) it appears that airfoils, the precursor to wings in true powered flight, first evolved as serial structures in both the fore- and hind limb. Most would agree that while dinosaurs like *Anchiornis* and the dromaeosaur *Microraptor* were incapable of the true powered flight that we see in the birds of today, it does not, however, exclude the possibility that they displayed some volant behaviour. The ability to flutter off the ground like a plump chicken in a barnyard, or to jump from trees, may have been the first step to acquiring actual powered flight.

One of the remarkable things about *Saurornithoides* and its close relative *Zanabazar* is their brains. Using CT scans it is now possible to determine what the brains of these animals looked like. Even though the brains themselves are not preserved, the brain case is. The interior surface of the braincase shows that the brains filled the entire back of the skull, and this means that if we digitally infill the brain case we can create an accurate representation of the size and shape of the brain. Analyses like these have been carried out on a number of different dinosaurs, and these data have shown that advanced dinosaurs had a brain size and complexity on a par with early birds.

Sometimes remarkable fossil animals are found that can tell us surprising things, and in exceptional cases this even includes behaviour. In the early 2000s, a remarkably preserved, small dinosaur was found in the Early Cretaceous Lujiatun beds in Northeastern China. While interpretations differ, many consider these beds to have been created by volcanic ash. Such volcanic clouds would also have contained noxious gases and falling ash

∧ The type specimen of *Saurornithoides*. Immediately Osborn recognized that it had implications for bird origins. This is reflected in its name which translates to lizard (*saur*) bird (*ornithoides*).

> A reconstruction of the small troodontid *Mei long* — the soundly sleeping dragon. It is depicted in the position it was fossilized in — curled into a ball with the tail wrapped around the perimeter and the head tucked between the arm and body.

would have accumulated quickly, as in the Roman cities of Pompeii and Herculaneum following the eruption of Vesuvius in 79CE. And so it was with *Mei long*, a 40cm (15½in) long troodontid specimen – the name *Mei long* in Mandarin means "soundly sleeping dragon". Not only is this a great specimen because of its completeness, it is also a great indicator of physiology.

The specimen was preserved sitting on its haunches, with its long tail wrapped around the perimeter of the body, and most notably, with its head tucked tightly under its arm. This is the same position that living birds adopt when resting or sleeping and is a way of decreasing surface area to maintain body heat. This is powerful evidence that *Mei long* was warm blooded.

All of these discoveries from other troodontids give us a new perspective on *Saurornithoides*. Osborn didn't know how prescient he was; if this animal were in a zoo today, to the casual observer it would look and act just like a strange bird.

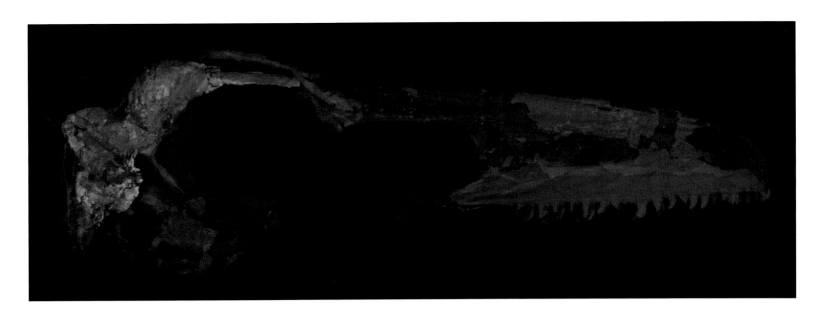

ARCHAEOPTERYX LITHOGRAPHICA

LATE JURASSIC
SOLNHOFEN FORMATION
GERMANY

ARCHAEOPTERYX (FROM THE GREEK FOR "ANCIENT WING") IS ONE OF THE MOST FAMOUS FOSSILS IN THE WORLD. THE FIRST SPECIMEN (JUST A FEATHER) WAS FOUND AROUND 1860 AND LATER DESCRIBED BY GERMAN PALEONTOLOGIST HERMANN VON MAYER, WHO COINED THE NAME.

Although there is some evidence that this is not the same species as the skeletons that were to come later, it holds an iconic position in the history of paleontology. In 1861 the first skeletal specimen was acquired by the Natural History Museum in London. The find, as noted by many illustrious paleontologists of the time, clearly showed that this was a strange creature. It had feathers, but also teeth, a long tail and fingers with claws. The presence of feathers – feathers that formed an airfoil, and a large fan-like tail – immediately

pointed to its relevance in the origin of birds, but in a 1974 paper, Yale paleontologist John Ostrom remarked that if the feathers had not been preserved, *Archaeopteryx* would have been considered just another small theropod dinosaur.

The Natural History Museum specimen (now called the London specimen) was immediately seized on by Charles Darwin and his adherents. *The Origin of Species* had been published a couple of years earlier in 1859. Some of the early criticisms of Darwin's ideas was that there were no intermediates in the fossil record.

Archaeopteryx was adopted by many of the early evolutionists as direct evidence of the "missing links" that define seminal points in evolutionary transitions.

As an evolutionary intermediate between traditional dinosaurs and birds, it is likely this is true for characteristics other than bones. Living birds grow fast; for example, the chickens we eat reach an age of 6–14 weeks old. On your next visit to the park, see if you can spot any baby pigeons – pigeons grow quickly and leave the nest in about a month. Non-avian dinosaurs grew more slowly. Although most were of larger size, they grew on a size-adjusted basis that was faster than crocodiles, but slower than birds. A study published in 2009 showed that *Archaeopteryx*, while not growing as fast as modern birds, grew faster than non-avian dinosaurs. Darwin was right that it was a transitional animal in most aspects.

< The first specimen of *Archaeopteryx* discovered. It came as quite a surprise to early paleontologists to find a modern type of feather in such ancient rocks.

> The amazing Berlin specimen of *Archaeopteryx*.

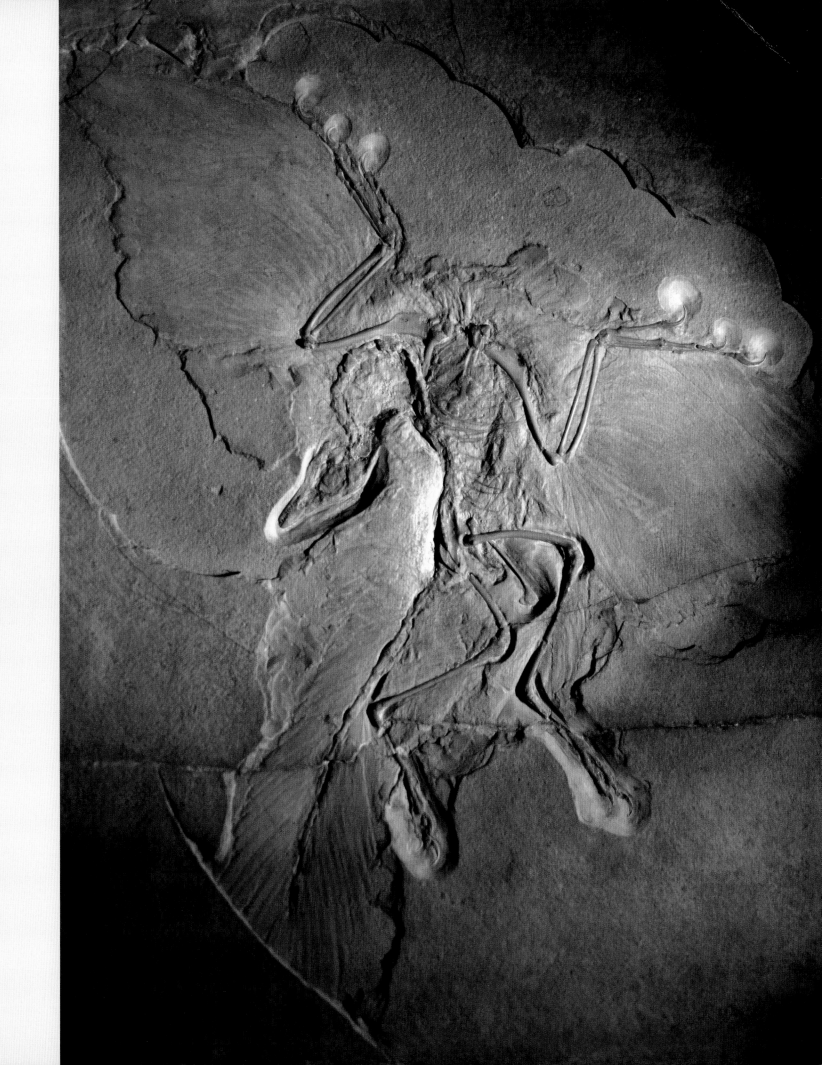

Archaeopteryx was a small animal, about the size of a crow. Because of its notability, it has been intensively studied on all levels. Much of this is contentious, but perhaps the most contentious area is whether or not it could fly. Although *Archaeopteryx* has a large airfoil on the forelimbs, and the forelimbs are covered with asymmetric feathers (usually associated with aerodynamic functionality), there are a number of parts of its skeleton that suggest it was incapable of true powered flight, at least in the sense of modern birds.

Flight, especially in animals that weigh over several hundred grammes, is hard and energy intensive. To compensate for this, birds have evolved a skeleton that is very rigid, which means that when they fly the entire energy in their muscles is transmitted to the flight stroke, not to kinetic movements in their body and tail. In living birds, much of the backbone is fused and the architecture of the shoulder girdle provides little opportunity for movement among the elements of the skeleton, which is optimized for transforming every single muscular flight stroke into the wings. They also lack a long, flexible tail. In contrast, the skeleton of *Archaeopteryx* is a loose one, and powered flight in the sense of the magical flight capabilities of modern birds was probably not possible, but what we call volant activity probably was. Consider a chicken fluttering through a barnyard to escape an intrusion rather than the flight of a swallow or swift when you think about this transitional aspect of flight evolution. True, modern bird-style flight probably did not evolve until the phylogenetic level of *Confuciusornis* in the Early Cretaceous. Thousands of specimens of this common proto-bird have been found in northeast China. Unlike more primitive forms, it has lost tail and teeth, has a rigid backbone and a firmly fused shoulder architecture. Undoubtedly it was capable of true powered flight.

> This is a fantastic specimen
of the small dromaeosaur
Microraptor. The feathers are
beautifully preserved. It was
this specimen that was sampled
to discover the coloration of
the animal.

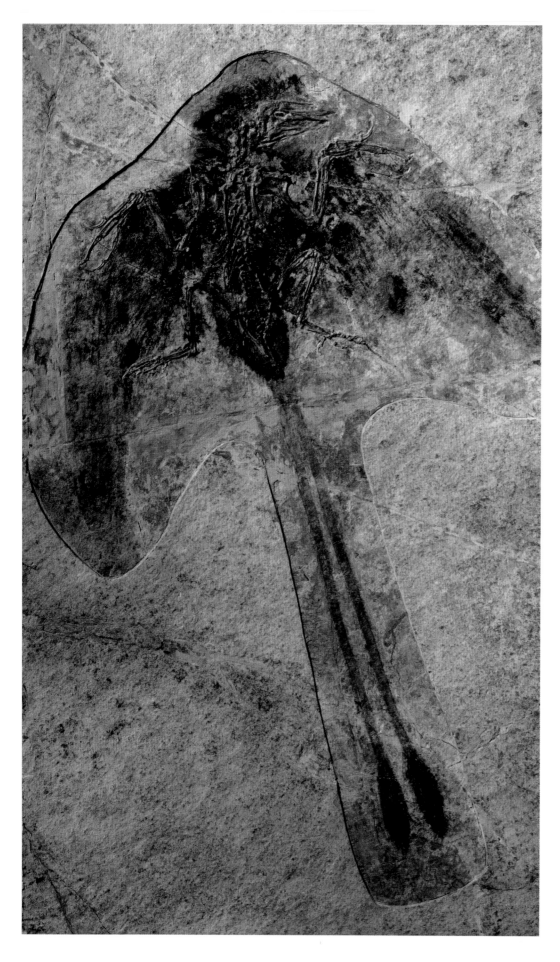

Archaeopteryx has been the subject of colour analysis. Colour in feathers is mediated by several elements. One is a class of compounds called carotenoids, which impart bright colours in birds such as the Northern Cardinal and the Scarlet Tanager, for example. However, structures called melanosomes are equally important, even if less exciting. Melanosomes are pigments containing organelles which exist inside cells, and which, almost unbelievably, can fossilize and be visualized using scanning electron microscopes after hundreds of millions of years. Following intensive mathematical analyses, it has been shown that melanosome shape is an indicator of what colour a feather used to be. Using this technique, it has been determined that the dromaeosaur *Microraptor* was flashy black, and the orientation of the melanosomes in parallel showed that it was iridescent. Such studies have been applied to a variety of dinosaurs from the rich Liaoning Early Cretaceous deposits.

These same techniques have been applied to *Archaeopteryx*. In 2011 detailed analyses of a purported *Archaeopteryx* feather indicated that the melanosomes were of a type that imparts the colour black. That is not to say that the entire animal was black, but this colour did form part of the plumage that existed in life.

The notoriety of *Archaeopteryx* has meant it has acted as a lightning rod for controversy. One example is bizarre claims by a number of famed astronomers and physicists during the 1980s that the specimens were forgeries. Although their motives remain unknown, one suggestion behind this claim is that British biologist Richard Owen, who never embraced Darwin's theory of evolution through descent with modification, wanted to set a trap to embarrass Darwin and Huxley. While firmly discredited, the popularity of dinosaurs continues to draw dilettantes to the field. Everyone from tech entrepreneurs, to acclaimed scientists in parallel fields, want to give opinions that, unfortunately, are usually poorly informed.

As an icon, *Archaeopteryx* remains a focus of all sorts of exciting science. Many more specimens have appeared in recent decades, and with each new specimen we appreciate that more questions will inevitably arise.

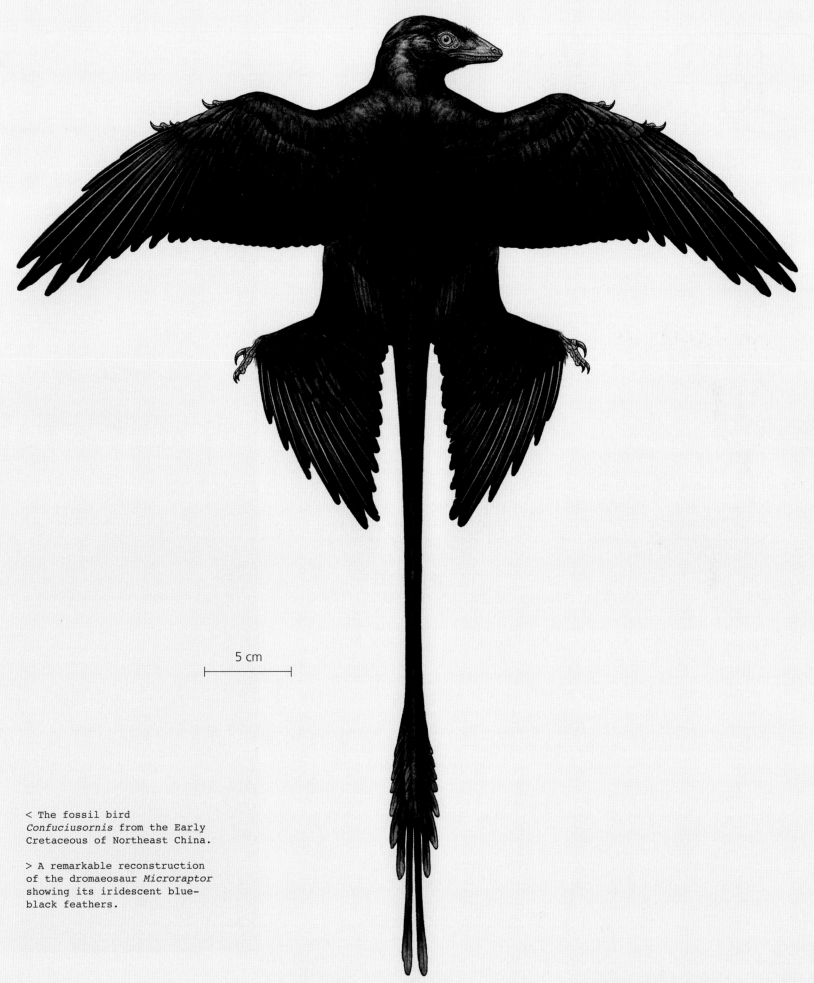

5 cm

< The fossil bird
Confuciusornis from the Early
Cretaceous of Northeast China.

> A remarkable reconstruction
of the dromaeosaur *Microraptor*
showing its iridescent blue-
black feathers.

HESPERORNIS REGALIS

LATE CRETACEOUS
NIOBRARA FORMATION
WESTERN NORTH AMERICA

BY THE MIDDLE OF THE CRETACEOUS MOST OF THE CONTINENTS HAD ASSUMED THEIR CURRENT CONFIGURATIONS – THAT IS, WITH THE EXCEPTION OF THE INDIAN SUB-CONTINENT, WHICH WAS A LARGE ISLAND IN THE PROTO-INDIAN OCEAN, SLOWLY DRIFTING TOWARDS AN EVENTUAL COLLISION WITH ASIA, MILLIONS OF YEARS AFTER THE EXTINCTION OF NON-AVIAN DINOSAURS.

Yet, even though North America was in the same position it lies today, it appeared very different. This is because it and other continents were inundated by water, which formed vast inland seas that teemed with life. Large, live-bearing seagoing lizards called mosasaurs reached nearly 15.2m (50ft) in length, and gigantic sea turtles and many species of large fish plied these warm, shallow waters. Pterosaurs of all sizes (some huge) flew through the skies. Ammonites, belemnites and other invertebrates swam in the seas. It was a unique environment that is unlike most contemporary marine habitats.

In addition to all this was an unusual bird. The term "bird" is used rather loosely here, as it is not a true bird, because all living birds are more closely related to one another than any of them are to this animal. Yet it is very closely related to the evolutionary radiation of modern birds, a close cousin to the living diversity. This animal is called *Hesperornis* and its remains (or those of very close relatives) have been found in all of the paleo-tropical seas of the northern hemisphere.

Hesperornis was a large, flightless bird, almost 1.8m (6ft) long. It was first discovered by the field crews of the famous Yale paleontologist O.C. Marsh in the early 1870s. What was so unusual about this bird was that, in addition to being flightless, it had teeth. These were restricted to the lower jaw, while the upper jaw showed characteristic signs that it supported a keratinous beak, just like birds today. It had many other features that indicates its close relationship with modern birds, and along with primitive characteristics like teeth, it was a poster child for evolutionary ideas documenting the origin of birds – so much so, in fact, that in 1873 Marsh declared: "The fortunate discovery of these interesting fossils does much to break down the old distinction between Birds and Reptiles." Over the decades the importance of *Hesperornis* as a transitional, yet very specialized, form has not diminished.

Hesperornis was a foot-propelled diver like a loon, in contrast to an underwater flyer like a penguin. Its forelimbs were much too feeble

> *Hesperornis* was like a toothed penguin and has many adaptations that point to its pisciverous lifestyle.

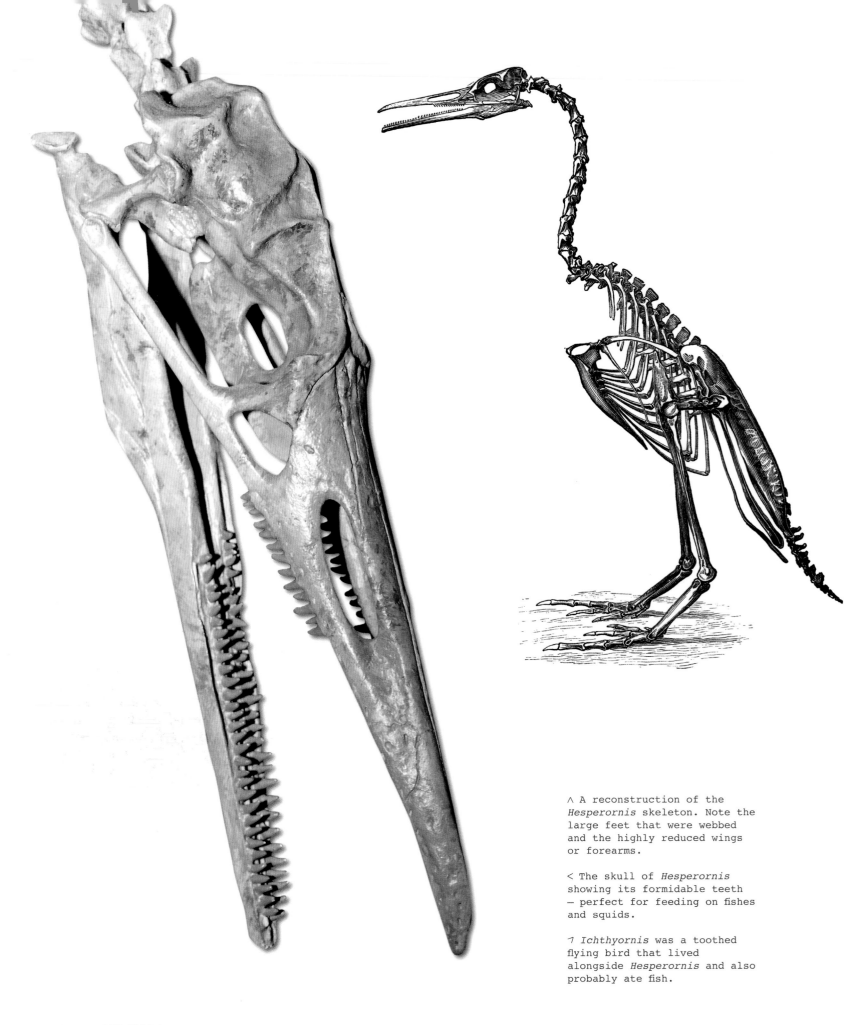

∧ A reconstruction of the *Hesperornis* skeleton. Note the large feet that were webbed and the highly reduced wings or forearms.

< The skull of *Hesperornis* showing its formidable teeth — perfect for feeding on fishes and squids.

⌐ *Ichthyornis* was a toothed flying bird that lived alongside *Hesperornis* and also probably ate fish.

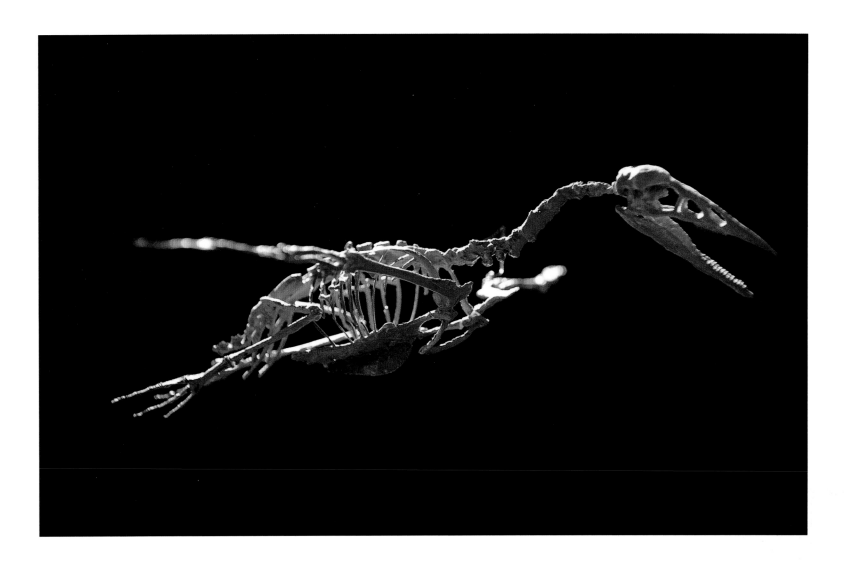

to be capable of use as water wings. Although we do not have any stomach contents, the long beak with teeth strongly suggests that it was a fish eater, which may have included squid and pelagic crustaceans. There is evidence that *Hesperornis* was occasionally prey itself: one specimen has been found that shows signs of attack by a plesiosaur (a seagoing reptile that was a common inhabitant of the interior seaway). This particular *Hesperornis* was lucky and escaped, as healing around the source of the tooth punctures is evident.

In the same Kansas sediments that produced the first *Hesperornis* specimens another, similarly toothed bird has been found. Named *Ichthyornis*, it was unearthed about the same time as the original *Hesperornis* finds, and since then its fossils have been found in most of the localities where fossils of the interior seaway have been recovered, from Saskatchewan to the Gulf Coast. This animal was much smaller, and it had teeth on both the upper (only in the middle

section) and lower jaws. Unlike *Hesperornis* it was at least a competent flyer and was probably the ecological equivalent of a modern-day seabird.

Ichthyornis played a prominent role in the so-called Bone Wars, a very public spat between America's two leading paleontologists, Yale's O.C. Marsh and E.D. Cope of Philadelphia. Benjamin Mudge of Kansas Agricultural College found many important fossils in the extensive chalk deposits of Western Kansas. As Mudge was not a paleontologist, he generously shared his discoveries with many specialists. He had enjoyed a close relationship with Cope for many years and had previously forwarded many important specimens to him. When Marsh found out about some of Mudge's new discoveries (including specimens of the toothed fossil bird that would come to be known as *Ichthyornis*), he interceded and encouraged Mudge to change the shipping address on the already packed crates. Therefore, instead of ending up in Philadelphia, the fossils were sent

to New Haven, and the spat became nasty.

Marsh first published on *Ichthyornis* in 1872. Yet, at the time, he completely misinterpreted what he had and felt that the toothed jaws belonged to a small variety of marine reptile associated with the skeleton of a bird. It was not until 1873 that he recognized instead that the jaw belonged to the skeleton of the bird itself.

Hesperornis was not known to have teeth until 1877, when good skull material was found. Hence, *Ichthyornis*, whose teeth were described by Marsh in 1873, was the first "bird" known to possess teeth at the time. Although *Archaeopteryx* had been previously described, the original specimen lacked the tooth-bearing elements of the skull (the first example to preserve the teeth was the Berlin specimen described in 1884). This bird with teeth was a discovery not lost on Darwin, who in 1880 proclaimed that the toothed birds from Kansas provided "the best support for the theory of evolution" in linking birds to their reptilian past.

PHORUSRHACOS LONGISSIMUS

MIDDLE MIOCENE
SANTA CRUZ FORMATION
PATAGONIA

IF YOU ARE FAMILIAR WITH THE 1961 DRAMATIZATION OF THE JULES VERNE SCIENCE FICTION CLASSIC CALLED *MYSTERIOUS ISLAND*, YOU WILL NO DOUBT REMEMBER THE SCENE IN WHICH A GIANT BIRD ATTACKS CONFEDERATE CASTAWAYS ON AN UNINHABITED TROPICAL ISLAND. A BIRD LIKE THIS – CALLED *PHORUSRHACOS* – DID INDEED EXIST IN THE MIOCENE EPOCH IN SOUTH AMERICA, ABOUT 13 MILLION YEARS AGO.

It was a large, flightless bird, standing around 2.5m (8¼ft) tall and weighing up to 130kg (287lb). The skulls of these animals were very large (up to 60cm/23½n) and were armed with a hook-shaped beak. Even bigger skulls of the closely related *Kelenken* (also from the mid-Miocene of Patagonia) measured up to 71cm (28in) in length, making them some of the largest avian skulls known. The head was mounted on a very stout neck, thought to be an adaptation for active carnivory. The feet also show predatory specializations in that they possess extremely large claws, and the legs indicate that the animal was capable of fast running.

A closely related species, *Titanis*, is also known from fragmentary remains in Florida and Texas, USA. The group is known to have a South American origin and the North American animals were descended from animals that took part in what is known as the "The Great American Interchange". This was an event that resulted from the formation of the Panamanian isthmus, which facilitated faunal migration between North and South

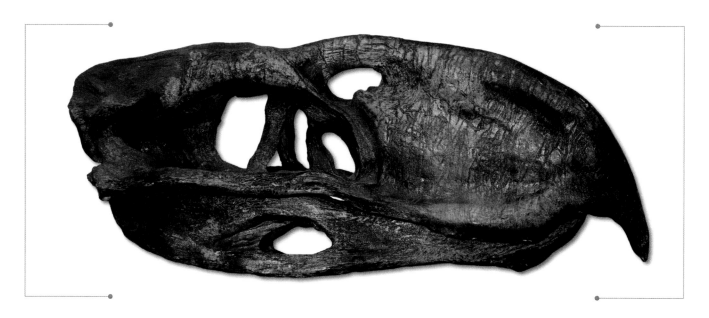

< The formidable hooked beak skull of *Phorusrhacos*. It is heavily built to resist the stresses of predation on large animals.

> A reconstruction of this predatory Cenozoic bird.

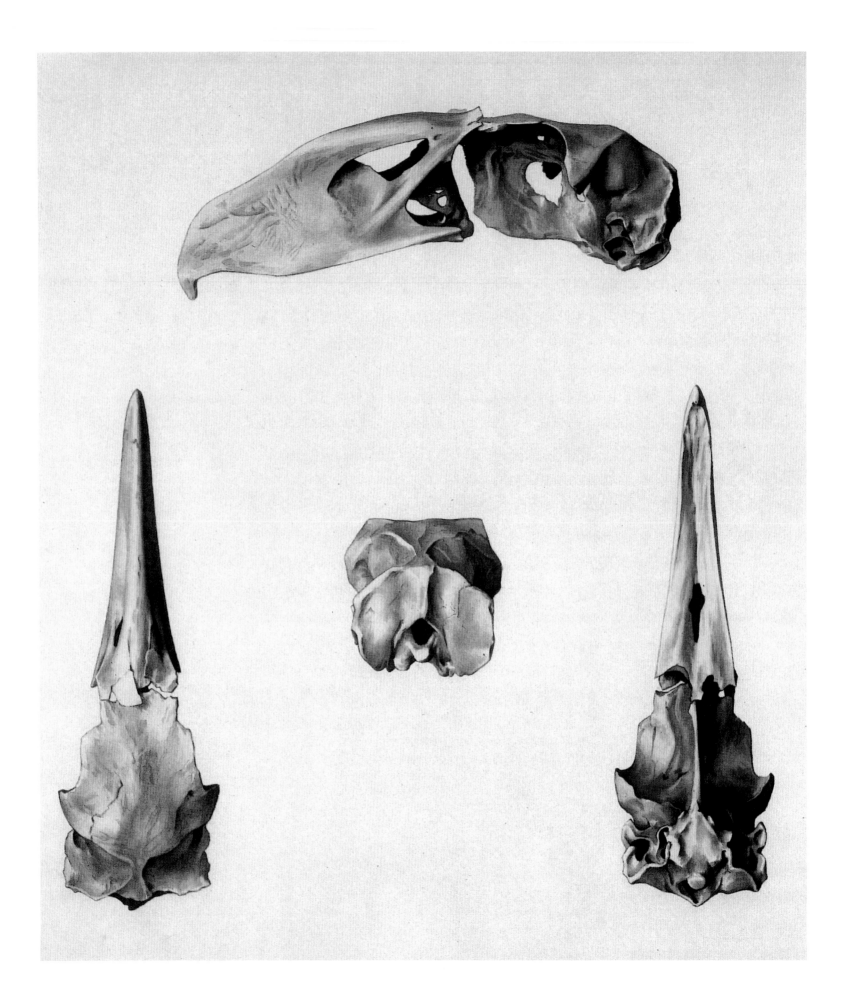

America. Among living animals, the closest relative of members of this group are the seriemas, a group of predominantly ground birds, which are highly carnivorous.

For much of its history South America was an island continent (much like Australia is today). Originally all of the continents were aligned into a single supercontinent called Pangea. The presence of this supercontinent explains why many early dinosaur faunas were so similar across the world. About 200 million years ago, Pangea began to break up. An important division was an east–west break that separated a large northern continent named Laurasia from a vast southern one called Gondwana. This lead to the development of more provincial faunas in each of the two continents. More fragmentation followed, and by about 100 million years ago, South America had separated from Africa. It evolved pretty much in isolation, producing unique and endemic dinosaurs. Some of these survived the Cretaceous–Paleogene mass extinction event about 66 million years ago and evolved into the endemic phorusrhacid group.

One of the peculiar things about the highly unique fauna of South America in the Cenozoic era, was it was largely depauperate in big mammalian predators. Instead, other animals such as large, terrestrial, lion-like crocodilians (for example, *Sebecus*) occupied these ecological niches. *Phorusrhacos* was also one of these top predators. Both *Phorusrhacos* and closely related forms have been posited as even being capable of bringing down large animals with their sharp beaks and long claws. Some have suggested that competition from large mammalian carnivorans (like felids and canids) which emigrated from North to South America may have played a key role in their extinction.

< A wonderful ink wash drawing of the skull of *Phorusrachus*. These kinds of illustrations were commonplace before the advent of modern digital methods.

> Charles R. Knight was one of the first of the early dinosaur illustrators. He worked closely with scientists and studied anatomy assiduously.

GASTORNIS GIGANTEA

LATE EOCENE
WASATCH FORMATION
WESTERN NORTH AMERICA

THERE ARE BIG BIRDS AND *BIG* BIRDS. *PHORUSRHACOS* (SEE P.130) WAS ANOTHER VERY LARGE FLIGHTLESS BIRD, BUT JUST BECAUSE AN ANIMAL IS BIG DOES NOT MAKE IT FEROCIOUS. A CASE IN POINT IS *GASTORNIS* (ALSO KNOWN AS *DIATRYMA*).

Fossils of this large bird were first found in Eocene deposits in Europe in the nineteenth century. Since then they have been found in North America and possibly in China. Early on, *Gastornis* received a great deal of notoriety; at over 2m (6½ft) in height, it was first thought to be a carnivore. And because it was found in sediments that were coeval with the early horse *Eohippus* (= *Hyracotherium*), one early reconstruction showed *Gastornis* with a struggling *Hyracotherium* in its mouth.

Yet there are several scientists who don't consider it to be a carnivore. Examination of the skeleton does not make it a very likely carnivore for a number of reasons. Its legs are elephantine and not really suited to the fast movement that a carnivore of this size would require. Furthermore, the head is extremely large compared to the body, and it is stoutly built rather than being light and flexible like most carnivores. Recent biomechanical analyses suggest that it had a powerful bite, one that is much more powerful than would be expected by a carnivore. Footprint evidence shows that *Gastornis* lacked the sharp claws usually associated with carnivorous dinosaurs (including birds). Finally, the chemical composition of the bones (especially calcium isotopes) is similar

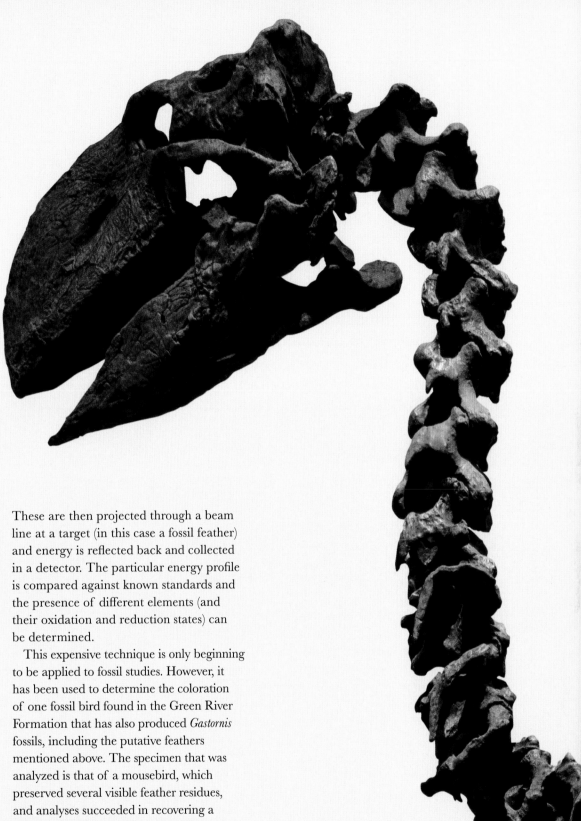

to that found in herbivorous dinosaurs and mammals and unlike the chemical make-up of unquestioned fossil carnivores like *Tyrannosaurus rex* and *Phorusrhacos*. Genealogical analyses have shown that *Gastornis* is related to ducks, geese and South American screamers.

What were presumed to be feathers attributed to *Gastornis* found in the Green River Formation, Colorado, have been shown to be the remains of plants. However, another feather has been attributed to this giant bird based primarily on its size (24cm/9½in in length). It is a broad-veined feather and is from the same locality. While analyses have yet to be conducted, such feathers can inform us about the animal's appearance using new techniques. Determining the grey, black and muddy red pigments using melanosome shape has previously been covered (see p.124) . However, another way to examine colour is to use powerful energy beams to detect the presence of minute amounts of colour-producing compounds. In this technique a synchrotron light source accelerates particles to a very high level.

These are then projected through a beam line at a target (in this case a fossil feather) and energy is reflected back and collected in a detector. The particular energy profile is compared against known standards and the presence of different elements (and their oxidation and reduction states) can be determined.

This expensive technique is only beginning to be applied to fossil studies. However, it has been used to determine the coloration of one fossil bird found in the Green River Formation that has also produced *Gastornis* fossils, including the putative feathers mentioned above. The specimen that was analyzed is that of a mousebird, which preserved several visible feather residues, and analyses succeeded in recovering a number of different elements. As these studies are only at a nascent stage, exactly how these elemental signatures translated into life coloration patterns has yet to be determined. Nevertheless, we no longer need to simply guess the colour of extinct feathered dinosaurs like *Gastornis* and *Microraptor* – with enough work and a little luck, we can figure it out.

PLATEOSAURUS ENGELHARDTI

LATE TRIASSIC
SEVERAL FORMATIONS
NORTH AND CENTRAL EUROPE

PLATEOSAURUS IS A VERY EARLY DINOSAUR, AND UNSURPRISINGLY DINOSAURS VERY SIMILAR TO IT EXIST ACROSS TODAY'S CONTINENTS, WHICH AT THE TIME WERE JOINED TOGETHER AS THE PANGEAN SUPERCONTINENT.

*P*lateosaurus is usually portrayed as a biped and more recent research confirms this. These were herbivorous animals that did very little oral processing of their food before swallowing it and, like the later giants to which these animals are related, may have had gut fermentation pouches. It has even been suggested that *Plateosaurus* may have supplemented its diet with carrion or small prey, making it omnivorous. Like many other dinosaurs, it reached adult size at around 20 years.

Plateosaurus fossils are abundant in the Swabian region of south-west Germany, where it is sometimes nicknamed the *Schwäbischer Lindwurm* ("Swabian dragon-monster"). Bones of what would become known as *Plateosaurus* were found as early as 1834, just slightly later than the recognition of the first dinosaur fossils in the UK.

For a long time, *Plateosaurus* and other similar dinosaurs were classified in the group *Prosauropoda* (meaning "primitive sauropods") with other apparently similar

dinosaurs. With the advent of new techniques to estimate genealogy, and with further research and more discoveries, the picture is far more complex, and the name *Prosauropoda* is no longer used.

The most primitive group of "prosauropods" is called the Plateosauridae, to which *Plateosaurus* belongs. The Plateosauridae were capable of bipedal locomotion, but other more advanced groups were less inclined.

However, many so-called "prosauropods" are more closely related to later sauropods than *Plateosaurus*; this is obvious from their anatomy. There are several of these, primarily from the UK, South America and southern Africa. These included animals like *Saturnalia*, a small 1.5m (5ft) long biped that is known from Argentina and perhaps Zimbabwe (due to the close proximity of these localities in Pangea). *Thecodontosaurus* is also similar and comes from the south of England; it is another of the first dinosaurs to be described (1836). Other "prosauropods", especially those from

Asia, reached large sizes. One of these, *Yunnanosaurus*, reached 7m (23ft) in length, and *Lufengosaurus* from a nearby locality, reached 9m (29½ft).

One thing that is remarkable about these "prosauropods" is that because they typically occur so early in dinosaur history (the Late Triassic and Early Jurassic), they are found on all continents (even Antarctica and Greenland) and are generally conservative in body plan. As the continents were all joined, extraordinarily similar animals occurred across the globe. For instance, one of these, *Anchisaurus*, is found in Connecticut in north-east USA, and a very similar dinosaur, which some consider to be synonymous, *Melanorosaurus*, is found in coeval rocks in South Africa.

The case of *Anchisaurus* is a curious one. It was one of the first dinosaur specimens to be discovered. Found in Connecticut in 1818, the discoverer thought the bones belonged to an ancient human. Other specimens were found in stone quarries in Massachusetts but were almost destroyed

by blasting. During the American Civil War a series of bridges was built over rivers in Connecticut. One of the workers noticed that some of the stones used in the construction were filled with bones. Part of this block containing a partial skeleton was given to the quarry owner and was later acquired by the Peabody Museum at Yale University. It was only in 1969, when the bridge was demolished, that the two sections of the skeleton were reunited.

⌐ *Plateosaurus* is one of the best examples we have of an early sauropodomorph dinosaur.

∧ Nothing is known of the body covering of *Plateosaurus*. What is known is that it was a semi quadruped.

> In some instances, several specimens of *Plateosaurus* have been found together in bone beds.

DIPLODOCUS LONGUS

DIPLODOCUS IS A SAUROPOD DINOSAUR. TOGETHER WITH *BRONTOSAURUS* (SEE P.144) IT WAS AMONG THE FIRST SAUROPOD DINOSAURS TO BE MOUNTED IN LARGE MUSEUMS.

Several species of *Diplodocus*, all from the same formation, have been named. The dinosaur became known internationally largely because so many casts of mounted skeletons have been donated to museums worldwide thanks to US industrialist Andrew Carnegie. Although *Diplodocus* had originally been discovered in 1877 and named by the Yale group in 1878, it was a specimen collected by Jacob Wortman and then described by John Bell Hatcher in 1901, which made *Diplodocus* such a cosmopolitan dinosaur. This specimen was extraordinarily complete and was destined for the nascent Carnegie Museum in Pittsburgh which was underwritten by the titan who had formed the United States Steel Corporation. Hatcher named the specimen *Diplodocus carnegii*. Andrew Carnegie was so thrilled, he made gifts of casts of the specimen to institutions around the world, including the Natural History Museum in London, where it was on display in the Central Hall until 2017. Casts were also donated to Berlin, Germany, St Petersburg, Russia, La Plata, Argentina, Bologna, Italy, and others.

Diplodocus belongs to a group of dinosaurs unsurprisingly known as Diplodocoidea which is found in many localities. It also includes some very diverse as well as closely related forms. *Amargasaurus* (from the Early Cretaceous La Amarga Formation of Patagonia), for instance, had a crest on its neck; the closely related form *Dicraeosaurus* from Tanzania also has a similar crest. Even one of the strangest and most exceptional sauropods, *Nigersaurus*, is a relative. Its jaws contained more than 500 teeth, and these may have been replaced as regularly as every 14 days.

⇗ This reconstruction of *Diplodocus* displays many of its salient features — a very long neck, a rather lightly built body and an extraordinarily long tail. Only recently has fossil evidence been found of the keratinous keels along the midline.

⌄ Barnum and Lillian Brown at the great dinosaur deposit of Howe Quarry in eastern Wyoming.

Among sauropod dinosaurs *Diplodocus* is one of the more common and well-known animals – even though there is disagreement over whether there has ever been a skull that can be definitively assigned to the type species. It is remarkable for several features in its skeleton. One of these is the position of the nostrils on top of the delicate skull, between the eyes. Some early paleontologists suggested that this was an aquatic adaptation, so the animal could feed with its head submerged and still breathe. While this is now discredited, more recently it has been suggested that these supported a trunk-like extension of the nose. Mammals today that have this sort of proboscis include elephants and tapirs, all of which have large narial openings high up on their skull. However, this view is not universally accepted.

A remarkable thing about *Diplodocus* is the length of the neck and tail. Cumulatively they make up most of the body length. These immense necks look as if they would have been disproportionally heavy, and had the weight of the tail not provided the correct balance, the animal would have constantly fallen forwards. But the neck is not as heavy as you might think: the individual neck bone segments (the vertebrae) are very light. CT scan analysis indicates the bony surface is very thin and surrounds large air spaces – in other words, the vertebrae are essentially hollow. But how could they be both hollow and strong? A very good analogy is the tower cranes that populate the skylines of most major cities around the world. These comprise lightweight but strong trusses composed of spans of steel. Due to their architecture they are rigid, but most of the weight is supported by very strong steel cables that extend from the end of the crane to a tower above where the operator sits.

This is just like in *Diplodocus* or another long-necked sauropod, but instead of a rigid frame of steel trusses surrounded by air, these sauropods have truss-like necks that are filled with air. And instead of taut steel cables carrying most of the weight, sauropods like *Diplodocus* have long tendons that attach to the neck bones and extend backwards to very high projections on the vertebrae just behind the neck. In this case, nature and human engineering converged on the same solution to construct a beam structure that is both strong and lightweight.

∧ Although not that closely related to *Diplodocus*, *Nigersaurus* displays many important sauropod features, like the retracted nose holes and copious teeth. It had around 500 teeth in its mouth.

< This cast of a *Diplodocus* skeleton used to grace the great hall of the Natural History Museum in London.

⌐ This *Amargasaurus*, from the Early Cretaceous of Argentina, displays a tall crest on its neck.

The tail of *Diplodocus* and many sauropods is also unique. As mentioned, it provided a counterweight to the animals' long necks. Early reconstructions of sauropods portrayed these animals dragging their massive tails on the ground. This popular image was adapted as the logo for the Sinclair Oil Corporation (see *Brontosaurus*, p.144) and is still used by them to this day. This view is now antiquated, as research over the last few decades clearly shows that these animals carried their tails aloft instead of dragging them, crocodile-like, along the ground.

One type of evidence is tracks. Dinosaur tracks are very common, and much can be learned from their study: how fast they moved and behaviour can be determined by looking at fossil footprints. Trackways of extant animals with long tails, such as Komodo dragons, large

iguanas and crocodilians, show a distinct and deep tail furrow between the hind limbs that represents a drag mark for the tail. No such marks are found in most sauropod trackways (or any other dinosaur for that matter). Clearly this indicates that the tails of these animals were carried parallel to the ground, not dragged along behind them. An analysis of the tail bones themselves shows that the individual vertebrae were supported, like the neck, along the top of the tail by strong ligaments emanating from high processes of the vertebrae at the hips.

Diplodocus is one of the few sauropod dinosaurs for which there is direct fossil evidence for soft tissue on the outside of the animal. Skin imprints have been found that show that the integument had a pliable, pebbly surface. It has even been suggested that these animals may have had a horny beak.

The end of the tail of *Diplodocus* is very strange. Some more recent discoveries at Howe Quarry, a famous site first excavated by Museum paleontologist Barnum Brown in the 1930s, have shown that the top of the tail was crowned with spikes, presumably formed of keratin, which stood about 18cm (7in) above the tail. At the end of the tail the vertebrae are tiny, 20cm (7¾in) long, and the thickness of a pencil. They lack strong processes needed for support and would have probably hung limply at the end of the tail, rocking back and forth as the animal walked. One interesting but untestable idea is that these animals used the ends of their tails as signalling devices: by whipping their tails, they may have been able to achieve supersonic speeds, creating a mini sonic boom like the whip of a lion tamer.

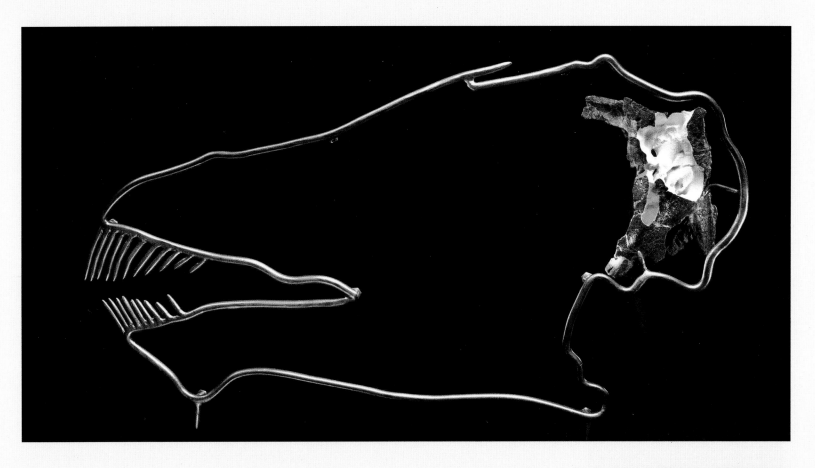

∧ The brains of even the most giant of these animals (here depicted in white) were very small.

∨ No "mummified" sauropod has been discovered, but tantalizing remnants of body coverings have been found like skin imprints and pieces of the keratinous fin.

> The extreme end of the tail of these animals was very delicate. Towards the tip each vertebral segment was only slightly more robust than a pencil, as in this *Diplodocus*.

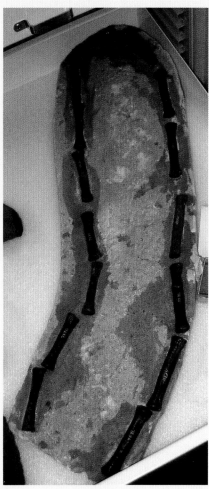

BRONTOSAURUS EXCELSUS

BELIEVE IT OR NOT, *BRONTOSAURUS*, WITH ITS UNUSUAL NAME MEANING "THUNDER LIZARD", IS ONE OF THE WORLD'S MOST BELOVED DINOSAURS. IT IS A LARGE SAUROPOD, ONE OF THE LARGEST FOUND IN NORTH AMERICA.

Its likeness has appeared across various media, from early animated films such as *Gertie the Dinosaur* (1914), to 1960s TV shows (*The Flintstones*), to oil company logos (Sinclair Oil Corporation), and it is one of the few (and most cherished) names known to the non-paleontological community. Like most other sauropods, *Brontosaurus* had a long neck, a stout body and a long tail. Also, like its close relatives, it had a relatively small, lightly built skull. Its teeth were restricted to the end of its snout and both its upper and lower jaws sported numerous pencil-shaped teeth. These teeth were not meant for chewing; rather, they were used for raking immense amounts of food into the mouth, perhaps up to several hundred kilograms per day.

Much of this food was of very low quality. When *Brontosaurus* lived, about 156 million years ago in what is now western North America, flowering plants or angiosperms had not yet diversified. Angiosperms are now the dominant plant group on Earth; they are rich in energy, and form the base of the food chain. We consume many directly – all of our vegetables and tubers – or indirectly because they form the main feed of the animals, such as chickens, cows and pigs, we eat. Garnering calories from

< Sculpting the head of the *Brontosaurus* for the Museum mount. Because no *Brontosaurus* skull had been found, one had to be fabricated. Unfortunately, they chose the wrong model, *Camarasaurus* instead of *Diplodocus*. This was rectified in the 1980s.

⌐ The Sinclair Oil Company was a sponsor for some of Barnum Brown's activities in the American West.

> Technicians assembling the *Brontosaurus* at the Museum in 1904. This was the first mount of a sauropod dinosaur ever constructed.

∧ The re-mounted *Brontosaurus*
in 1995 as it was installed in
the saurischian dinosaur hall
at the Museum. In a departure
from the previous pose, the
head only slightly elevates
above the level of the body
and the tail is held high off
the ground.

such poor-quality food is difficult. We do not know exactly how the giant sauropods digested their food, but several reasonable ideas have been proposed.

The most popular of these is that the animals harvested plant fodder and stored it in vast outpocketings, or diverticula, in the digestive tract. There the food sat and fermented and the energy that was absorbed by the dinosaur was the fermentation products of the bacterially digesting plants. A similar mechanism has evolved in the Galápagos tortoise, which lives today on a small island archipelago in the Pacific Ocean. While we cannot tell for certain that this was the mechanism for digestion, this is the most reasonable explanation that has yet been put forward.

The nomenclatural history of *Brontosaurus* is complex. Originally for the animal two genera of dinosaurs were confused, and this is just beginning to be resolved. Scientific names are regulated by the rules of scientific nomenclature, which is a policy that outlines

the ways in which scientists can affix names to organisms. One of the tenets of this policy is that when a species is named it must be unique, and the first name given to an animal is the valid one.

In the early 1870s, Yale University field crews sent to the American West found many great specimens. One of these was described as *Apatosaurus* by Yale paleontologist O.C. Marsh in 1877. Two years later, in 1879, another specimen was found and described by Marsh. He named this specimen *Brontosaurus*. Both

specimens were rather incomplete. By the early 1900s it was postulated that these two animals represented the same species and in deference to the rules of scientific nomenclature, and since *Apatosaurus* was used first in the scientific literature, it was regarded as the valid name. This was amplified in the early 1990s, when extensive work was done on these fossils and the name *Brontosaurus* was relegated to the dustbin, even though it is far more familiar to the general public than the rather obscure *Apatosaurus*. Old habits die hard, and many

who had grown up with *Brontosaurus* as a dinosaur icon in popular culture complained vociferously, but to no avail.

However, since 2015, according to a re-examination of these animals, *Brontosaurus* has now been rehabilitated. These new studies provide powerful evidence that *Brontosaurus* and *Apatosaurus* are distinctly different and both valid names for two different kinds of large sauropod dinosaurs. Yet the argument is far from settled and the validity of the name *Brontosaurus* is still embroiled in controversy.

CAMARASAURUS GRANDIS

LATE JURASSIC
MORRISON FORMATION
WESTERN NORTH AMERICA

DURING THE LATE JURASSIC IN NORTH AMERICA, WHEN SAUROPODS REIGNED SUPREME, *CAMARASAURUS* WAS THE MOST COMMON OF THEM ALL.

Its fossils are found throughout Morrison Formation localities. Adults were big, up to 23m (75½ft) in length. The name *Camarasaurus* means "chambered lizard", which refers to the large cavities and holes in the vertebrae of the animals. This is a common feature of giant sauropods and was surely a weight-saving measure that may also have had some respiratory consequences.

Camarasaurus was quite different from other Morrison sauropods such as *Brontosaurus*, *Barosaurus*, and *Diplodocus*. Instead of a long, low head, it had a very blunt snout and a highly domed skull with enormous nasal openings. These have been compared to those of elephants and tapirs, suggesting that the front of the face may have supported a trunk-like structure. However, this idea is not held in high regard. *Camarasaurus* had a long neck that appears to have been held higher above the ground, and in a more vertical position than many other sauropods, in a similar way to *Brachiosaurus*. Its teeth were also different: instead of the thin, pencil-like teeth found in *Diplodocus* and *Brontosaurus*, the teeth of *Camarasaurus* were bulky and spoon-

> An exceptionally complete juvenile *Camarasaurus* skeleton was discovered in a typical "death pose" at Dinosaur National Monument, Utah, USA.

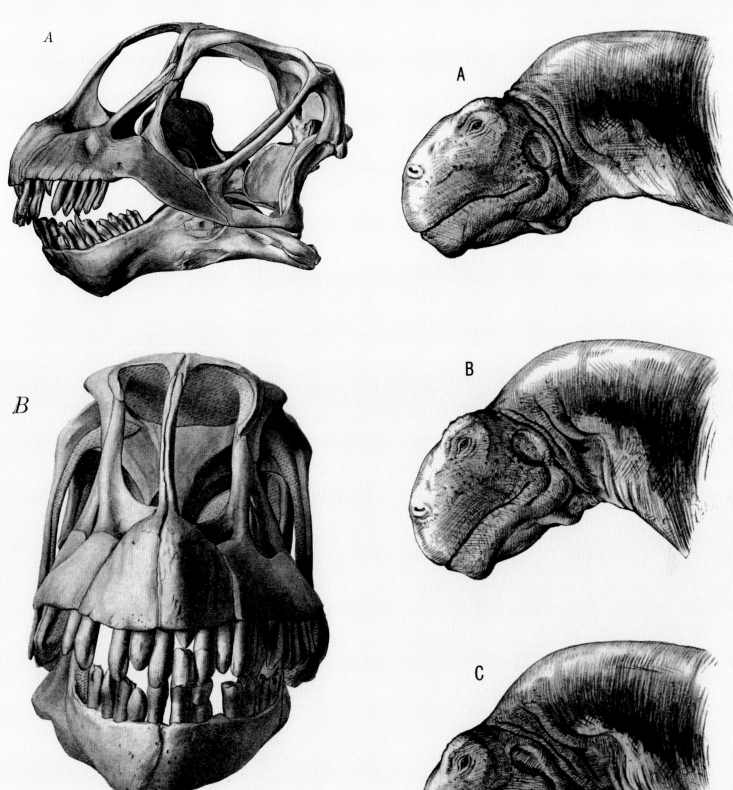

∧ The skull of *Camarasaurus* is high domed, and the teeth are massive, unlike the condition in *Brontosaurus* and *Diplodocus*, where the teeth are pencil-like and the skulls are low in profile.

> This rendering of the face of *Camarasaurus* by Erwin Christman gives the animal an almost gestural personality.

shaped and often show strong wear facets. This suggests that these animals may have been capable of more oral processing of their fodder than many other large sauropods.

In an interesting twist, it was the skull of *Camarasaurus* that long adorned the *Brontosaurus* at the American Museum of Natural History. The original *Brontosaurus* specimens had all been found without skulls. For decades the skull of *Brontosaurus* was unknown, and so when the first *Brontosaurus* was mounted at the Museum in 1905, the skull was reconstructed to look like a large *Camarasaurus*, because at the time *Brontosaurus* was thought to be closely related to *Camarasaurus*. Research during the 1980s showed that actually *Diplodocus* and *Brontosaurus* were closely related and this led to the identification of several *Brontosaurus* skulls all looking like *Diplodocus*.

Camarasaurus is the most common dinosaur of Wyoming's Howe Quarry, but unfortunately most specimens were destroyed through neglect in the lean years of the Great Depression and the Second World War. Operations at Howe Quarry were renewed in the 1980s by private commercial fossil collectors, who found many *Camarasaurus* specimens. Among these remains were elements of soft tissue; some of these can be definitely tied to *Camarasaurus*, while others are probably *Diplodocus*, or even one of the rarer Morrison sauropods. The *Camarasaurus* soft tissue is best preserved around the jaws, which indicates that the teeth were deeply set in fleshy gums with only their tips protruding. The other Howe Quarry soft tissues that are probably from *Diplodocus* include long, triangular keratinous spikes on the top of the tail and possibly the back. It is

interesting that so many soft tissue discoveries have been found in recent years at localities that were first sampled decades ago. Whether early paleontologists were not looking for these materials or they were destroyed during preparation is not known. Their discovery has, however, dramatically changed the way we perceive these animals.

> During the time that the Howe Quarry excavation was active it became quite a tourist attraction.

v The skull and neck of a *Camarasaurus* specimen are displayed as discovered in the Late Jurassic rocks at Dinosaur National Monument.

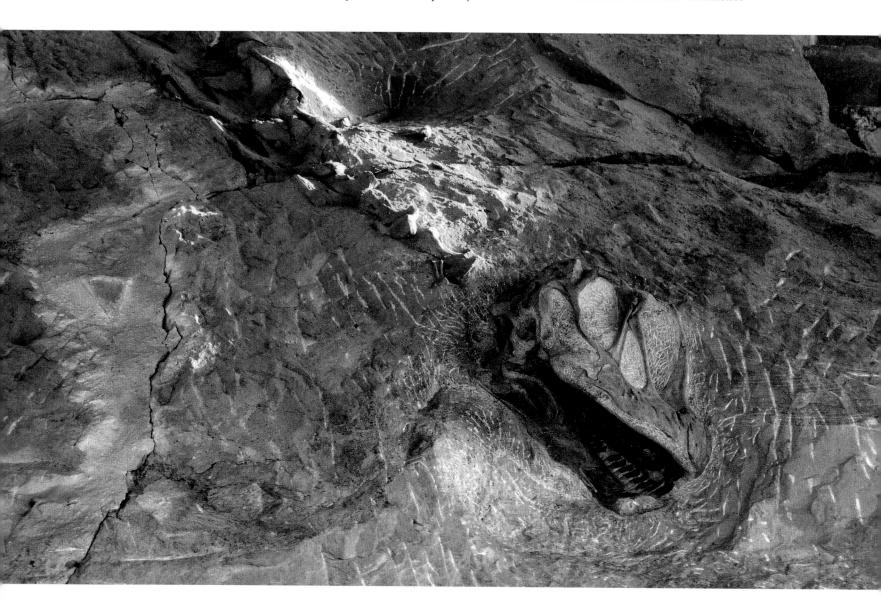

HYPSELOSAURUS PRISCUS

LATE CRETACEOUS
GRÈS À REPTILES FORMATION
SOUTHWEST EUROPE

WHEN WE THINK OF DINOSAURS, WE USUALLY DON'T THINK OF EUROPE AS A HOTBED OF PREHISTORIC ACTIVITY, EVEN THOUGH IT IS WHERE THE FIRST DINOSAURS WERE FOUND AND IDENTIFIED AS THE BONES OF ANCIENT REPTILES. AND WITHIN EUROPE THE IDYLLIC COUNTRYSIDE OF THE SOUTH OF FRANCE IS MORE READILY ASSOCIATED WITH FINE WINE, FOOD AND IMPRESSIONIST PAINTERS THAN DINOSAUR PALEONTOLOGISTS.

But in 1846, shortly after dinosaurs had first been described in England, a French geologist, Pierre Matheron, described the first remains of this sauropod dinosaur from rocks interspersed with the fine vineyards of Provence.

Through the decades, and with more discoveries, these remains were recognized as belonging to the titanosaur group, the most ubiquitous group of sauropod dinosaurs worldwide in the Late Cretaceous. This group includes titanosaur immigrants to North America late in the Cretaceous such as *Alamosaurus* from Texas; peculiar, long-necked varieties from Asia like *Erketu*; Patagonian and Asian giants; and strange African and Indian forms which, armoured and with tail clubs, is unusual for sauropods. This group even includes dwarf forms such as *Europasaurus* from Europe, which was an island archipelago in the Late Cretaceous. Instead of giant behemoths, these animals reached adult size at just over 6m (19½ft) and have been used as an example of island dwarfism.

One of the most unique things about *Hypselosaurus* is that when Matheron

was excavating further specimens after the initial remains had been found, he came across fragments of large eggs. Examining the circumference of the eggs, he postulated that they were too small to have been the eggs of *Hypselosaurus*, which he correctly deduced, based on the length of the femur, was a very large sauropod at least 15m (49¼ft) in length. He based his approximation on eggs that were known at the time from the elephant bird (*Aepyornis*) recently extinct from Madagascar. It turns out that Matheron was incorrect and that *Aepyornis* eggs are disproportionately large (compared to body size). While there is no unequivocal evidence that these are *Hypselosaurus* eggs, there is very good evidence that they are titanosaur eggs. The eggs were found in association with *Hypselosaurus* bones, but no embryo has

> Not too much is known about the skeleton of *Hypselosaurus*. Consequently, renderings need to be heavily based on closely related animals.

been found inside one of the eggs to make the identification certain.

Many other definitive titanosaur eggs have been found in other parts of the world, and detailed examination of these has allowed us to identify titanosaur eggs that do not contain embryos. The most spectacular of these comes from a locality in Argentina called Auca Mahuevo, where large associations of titanosaur eggs suggest that they may have nested communally. There are a number of peculiar points to note about these Argentine eggs. One is that many preserve embryos; although embryonic dinosaur fossils have become more common in the last couple of decades, these are special. In addition to preserving the tiny bones and teeth, these featured fossilized skin, thus giving us our first impression of what baby sauropod skin looked like. Other things about some Argentine sauropod nesting grounds are also fascinating. Perhaps the most unusual is the proposition that the eggs may have been "incubated" by hydrothermal vents. Most current ideas suggest that the majority of dinosaur eggs (except those most closely related to birds) were buried in nests of vegetation and sand that stabilized their humidity and temperature in the same way as living crocodiles. As strange as the hydrothermal vent hypothesis sounds, a similar behaviour is found today in some species of megapode birds, which nest near volcanic vents and use geothermal heat to incubate their eggs.

Titanosaur nests are known throughout the world and have provided an insight into how these animals lived. In India, they have been found associated with the skeletons of large snakes in the nests; whether these were chance occurrences or whether the snakes were preying on eggs or young, is not known. In Eastern Europe titanosaur nests are common, and are often linear, where the egg-layer has apparently laid the eggs in a row, then, presumably, has covered them. All that we really know is that the diversity and reproductive habits of these animals, considering their differences in ecology, diet, size and distribution, could easily eclipse what we see in the living dinosaurs of today.

∧ *Erketu* was a relatively small dinosaur. However, proportionally it had the longest neck of any known dinosaur.

∨ Although the skeletons are rare, the eggs of *Hypselosaurus* are ubiquitous in southern France. Sometimes great accumulations of these eggs have been found together, perhaps indicating communal nesting.

> Many reconstructions of sauropod nesting sites depict them as communal nesters. Here a *Diplodocus* tends its young on a floodplain in what would be the American West about 154 million years ago.

PATAGOTITAN MAYORUM

LATE CRETACEOUS
CERRO BARCINO FORMATION
SOUTHERN SOUTH AMERICA

WHAT WERE THE BIGGEST DINOSAURS? THIS IS A DIFFICULT QUESTION BECAUSE THERE ARE A NUMBER OF CONTENDERS. MANY AROUND THE PLANET ALL COMPARE IN TERMS OF SIZE BUT SEVERAL ISSUES CONFOUND THE CHOICE OF A CHAMPION; IS BIGGEST THE TALLEST, THE HEAVIEST OR THE LONGEST? THE QUESTION IS LIKELY TO GENERATE THREE DIFFERENT ANSWERS. ALL THINGS BEING EQUAL, *PATAGOTITAN MAYORUM* IS ONE OF THE LARGEST, IF NOT THE LARGEST, EVER COLLECTED.

In 2008, a worker at an estância in southern Argentine Patagonia, came across some bones weathering out of a small hill in a very arid valley. He reported these to the owner, who then alerted paleontologists at the regional museum in Trelew, the capital of Chubut province. The abundance of fantastic fossils in this area has made it a hotbed of paleontological discovery for over 100 years. This has fostered a very talented group of paleontologists at the Egidio Feruglio Paleontology Museum in Trelew. When they first saw some of the specimens, they were immediately enthusiastic: this was a very large animal and competitor for the largest land animal to exist.

Excavations in the approximately 100-million-year-old sediments soon began. As a lot of soil had to be removed, and due to the immensity of the bones, progress took place over several excavation seasons. More than 150 bones of six different individuals were collected, and what they found was one of the most dramatic animals to be exhumed anywhere on the planet.

Analysis of *Patagotitan* has demonstrated that it measured about 40m (131ft) in length, and was 5.5m (18ft) high, although others have revised the length slightly downwards. The skeleton was so large that a single femur (upper leg bone) measured over 2m (6½ft) in length. The fossilized specimen weighed over 400kg (882lb). Alive, *Patagotitan* has been estimated by some to have weighed around 69 tonnes as an adult; its neck is estimated at over 12m (39⅓ft) in length – and the type specimen is only about 85 per cent grown. When alive, *Patagotitan* roamed a forested area that was bisected by

> *Patagotitan* on display at the Field Museum in Chicago.

INFORMATION

< This reconstruction shows a rather drab *Patagotitan* that is almost elephantine. In life there was probably considerably more colour and variability in the skin texture over the entire body.

> Many titanosaurs, the group to which *Patagotitan* belongs, have large bones embedded in the skin (called osteoderms) and even large clubs at the end of their tails. This titanosaur osteoderm from India is as large as a dinner plate.

meandering streams, much like parts of East Africa today.

Is *Patagotitan* the largest dinosaur ever found? There are a few other contenders. One of these is an animal named *Argentinosaurus*. Also from Patagonia, this was a massive animal, comparable in size to *Patagotitan*. However, very little of the *Argentinosaurus* skeleton is known, so direct comparisons are difficult. Similarly, a highly fragmentary skeleton named *Ruyangosaurus* has been found in Szechuan province in China, in rocks that are about

115 million years old. Discovered during road construction, it is also fragmentary and many of the preserved bones have yet to be exhumed.

All of these giants belong to the same group of dinosaurs, the Titanosauria. Titanosaurs are a ubiquitous group in the Cretaceous, and their remains have been found on all continents. They are, however, considered to have their origin in Gondwana (see p.32-33), although there are a few Laurasian examples in North America and Eurasia. Their body plan was pretty much stereotypical for

sauropods (long neck, long tail, small head and big body), with a couple of caveats. Some titanosaurs had body armour consisting of lumps of bone embedded in their skin. This body armour, called osteoderms, is fairly common in many ornithischian dinosaurs, but it is not present in other sauropod groups. Also, mimicking the ornithischian group Ankylosauria, some titanosaurs had large bony masses or clubs at the end of their tails; we can only speculate what these might have been used for.

BAROSAURUS LENTUS

LATE JURASSIC
MORRISON FORMATION
WESTERN NORTH AMERICA

THE MORRISON FORMATION IS A SEQUENCE OF LATE JURASSIC ROCKS THAT IS EXPOSED IN VAST AREAS THROUGH THE INTERMOUNTAIN AMERICAN WEST. THESE SEDIMENTS PRODUCED MANY OF THE HISTORIC DINOSAURS ASSOCIATED WITH THE EARLY PHASES OF NORTH AMERICAN DINOSAUR COLLECTING.

Many of these, like *Allosaurus* (see p.64) and *Bronotsaurus* (see p.144) have already been described. Collected in the late nineteenth century, many of these large dinosaurs were soon mounted in the early twenteith century in museums around the world. One of the rarest of the large Morrison sauropod dinosaurs is *Barosaurus*.

Barosaurus was a lightly built but extremely elongate sauropod dinosaur. Adults reached 26m (85¼ft), but there is some indication that animals almost twice as large may have been present. *Barosaurus* differs from closely related Morrison animals in proportion: it has a longer neck and a shorter tail than *Diplodocus*, and a more lightly built skeleton than *Brontosaurus*.

One of the characteristic features of sauropod dinosaurs is their extremely long necks. This was taken to a great extreme in the titanosaurid *Erketu* (see p.154) collected by scientists of the Mongolian Academy of Sciences and American Museum of Natural History in the Eastern Gobi locality of Bor Guve in Mongolia. Although it is not an extremely large dinosaur, its neck is probably proportionally longer than any other known sauropod relative to its body size. Many other large dinosaurs, such as *Omeisaurus* and *Mamenchisaurus*, also had proportionally very long necks.

Ever since they were discovered the use and position of these necks have been the subject of debate. One early idea was that the animals were aquatic. Many sauropod nostrils are on the tops of the animals' heads right in front of the eyes, and it was proposed that this meant they could breathe through these openings while their heads were submerged feeding underwater. This theory is easily dismissed due to the issue of buoyancy. The bodies of sauropods, like other dinosaurs and especially saurischians, were not very dense, and would have floated easily. The most common idea is that long-necked sauropods used their necks like giraffes to pick vegetation off tall trees. This sounds reasonable, but on close inspection this also becomes untenable.

The individual segments of the backbone of *Barosaurus*, including the neck and tail, are

(except for the very tip of the tail) connected to one another by flat processes that allow movement. The planes of movement are specifically determined by the orientation of these surfaces relative to one another. In most sauropod dinosaurs analysis of these surfaces indicates that very little vertical movement of the neck was possible. These animals were incapable of lifting their heads much above the height of their backs. So, the question remains, if you can't nibble at high foliage why evolve a long neck?

The answer is intuitive. Sauropod dinosaurs were the largest animals ever to walk on Earth, both in volume and in mass – although their mass has probably been overestimated. Even if they had a sluggish metabolism, it would have taken a huge amount of energy to sustain bodily function at their weights. As an example, living African elephants weighing a fraction of what the largest sauropods would have weighed, feed up to 18 hours a day and consume about 300kg (661lb) a day in fodder. This is relatively high-calorie fodder, mostly composed of angiosperms, which were not the main diet of most sauropods because angiosperms were not sufficiently diversified until well after most of the largest dinosaurs had become extinct.

Sauropods had to sustain themselves with tree ferns, cycads and conifers, which are relatively low-quality foods. One theory that has been suggested is that the sauropod digestive track contained large outpocketings which acted as fermentation tanks, where plant material was broken down with the aid of bacteria into fermentation products, which are both easy to digest and higher in calories. If so, in addition to being huge, sauropods would also have been particularly flatulent.

It has been calculated that sauropods as large as *Barosaurus* would have needed to consume hundreds of kilograms of available food per day to flourish. If you are going to consume food at this rate, it is advantageous to conserve as much energy as possible while eating. One way to do this is to have a very long neck that can sweep back and forth and accumulate as much food as possible without taking a step forward. The area over which the animal feeds is called the feeding envelope; the longer the neck, the larger the feeding envelope and the greater the conserved energy.

⌐ A vertebra from *Barosaurus*. Although it looks massive, the bone is actually quite light. The surface is very thin and large air spaces occupy the interior.

< The rugged canyons of eastern Utah display superb exposures of the Morrison and other Mesozoic Formations. This is where the Museum's *Barosaurus* was collected.

∧ In 1993 the Museum mounted a cast of its *Barosaurus* specimen in a controversial pose. It is portrayed rearing up in defence of its young in front of a surging *Allosaurus*.

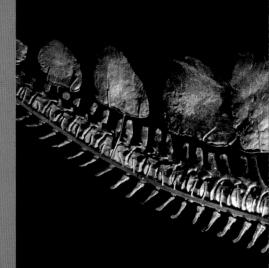

ORNITHISCHIA

THYREOPHORA
168

NEORNITHISCHIA

MARGINOCEPHALIA
180

ORNITHOPODA
206

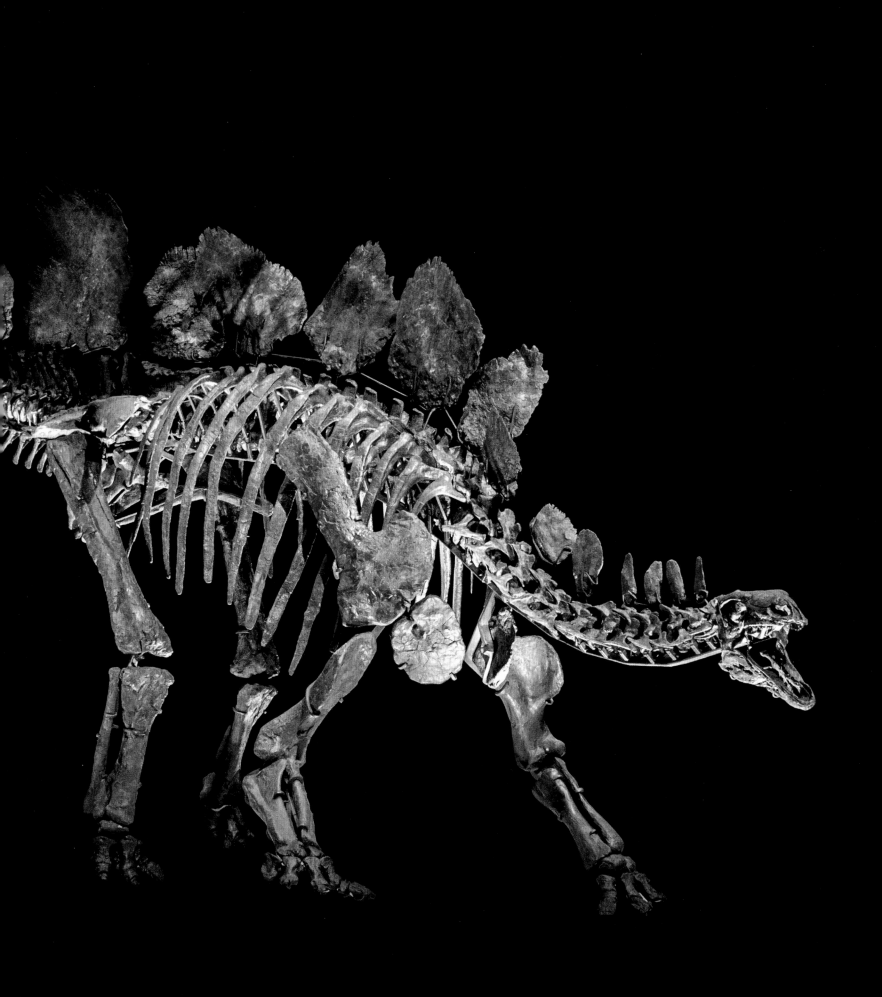

HETERODONTOSAURUS TUCKI

EARLY JURASSIC
ELLIOT FORMATION
AFRICA

HETERODONTOSAURUS WAS A SMALL (ABOUT 1.5M/5FT), BIPEDAL AND PRIMITIVE ORNITHISCHIAN DINOSAUR. AS THE NAME IMPLIES, ITS MOUTH CONTAINED DIFFERENT KINDS OF TEETH: SMALL, INCISOR-LIKE TEETH AT THE FRONT BEHIND THE TOOTHLESS HORNY ENDS OF THE JAWS, LARGE CANINES, AND CHISEL-LIKE CHEEK TEETH.

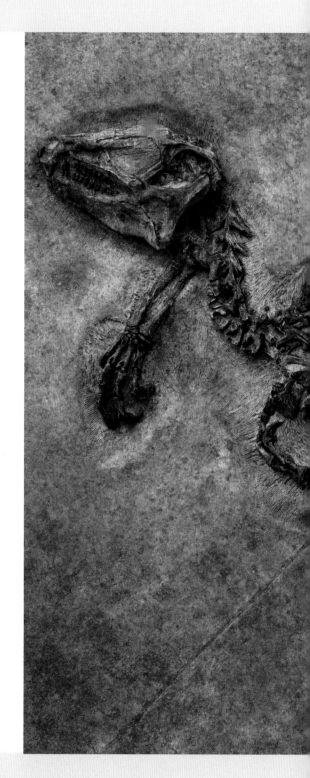

Unlike most ornithischians the cheek teeth were of very different sizes, with the largest located in the middle section of the tooth row. Several possibilities of the relationship of these teeth to the diet of *Heterodontosaurus* have been proposed. Both herbivory and omnivorey have been suggested, as it is not just the size of the teeth that varies along the tooth row; the large canines, for instance, show small serrations on their back edges. One hypothesis is that these tusks were used for secondary sexual display in combat against rivals in competition for mates, but this has been somewhat contradicted as nearly all (although there are not many) *Heterodontosaurus* specimens had tusks. It is perhaps more likely that they used these tusks, like peccaries or warthogs, to forage for subterranean foodstuffs.

Although *Heterodontosaurus* was a very primitive dinosaur, the discovery of a close relative in 2010 makes it more relevant. Named *Tianyulong*, the find was made in the Early Cretaceous sediments of the Jehol Group in Northeastern China, which have produced many important dinosaurs in recent years. Soon after it was discovered,

additional specimens started to appear. Like *Heterodontosaurus* it was a small animal – even more diminutive at 70cm (27½in). What is remarkable about these specimens is that they were covered with small filaments, which occurred in patches along the back and tail, and the top and bottom of the neck; other specimens suggest that the filaments covered the entire body. On close inspection the filaments are very similar to the structures found on the theropod dinosaur *Sinosauropteryx* (see p.82) and the feathered therizinosaur *Beipiaosaurus*, also from the Jehol Group. In all of these specimens the structures consist of single, unbranching filaments.

This has important implications for the occurrence of feathers in dinosaurs. Because *Tianyulong* is a close relative of *Heterodontosaurus* near the base of the dinosaur family tree, it indicates that all dinosaurs had feathery structures, and that such primitive feathers were present in the common ancestor of all dinosaurs.

The story may be even more complex. It has been known for several years that some pterosaurs (flying reptiles) are among the closest relatives of dinosaurs. It has also

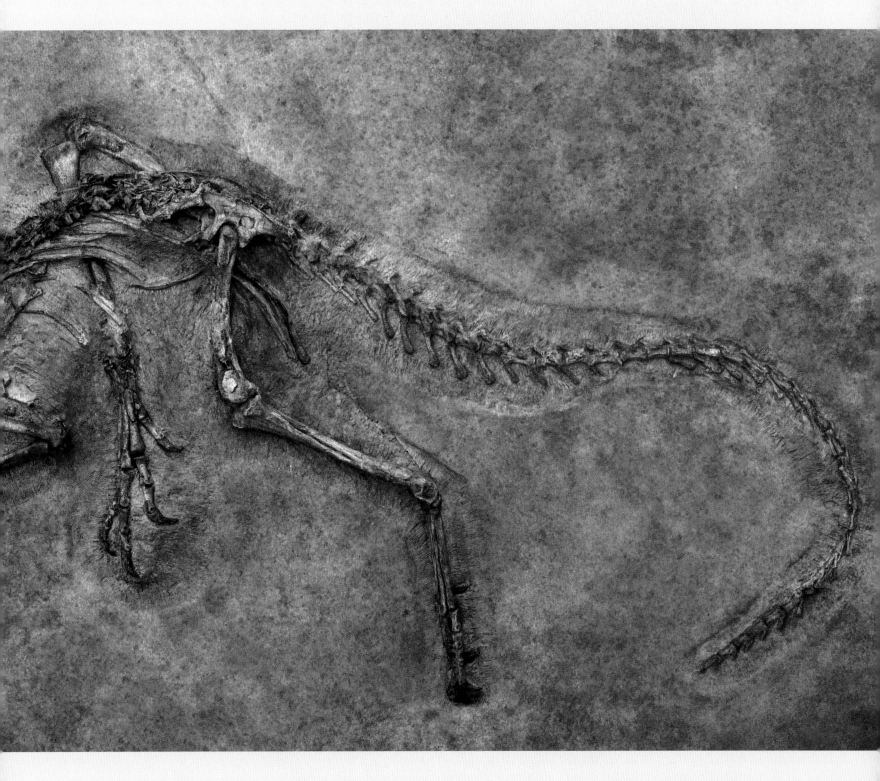

∨ The type specimen of
one of the most primitive
ornithischian dinosaurs known,
Heterodontosaurus.

< A reconstruction of *Heterodontosaurus*.

∟ A close relative of *Heterodontosaurus* is *Tianyulong*. *Tianyulong* is from the early part of the Late Jurassic of China. Fossils show that it had primitive feathers all over its body.

∨ The holotype specimen of *Tianyulong*. The black marks on the upper left are feather impressions.

been known that pterosaurs, like dinosaurs, were covered with structures that resemble the filaments on *Sinosauropteryx* and *Tianyulong*. This indicates that these structures probably evolved separately from dinosaurs, at least in the ancestor of pterosaurs and dinosaurs, but how far could it be pushed down the archosaur tree? The ancestral archosaur (the species that gave rise to both crocodiles and dinosaurs, including birds), was a small, energetic biped. Several of the animals along the progression that leads to modern types of crocodile look more like primitive, dinosaur-line archosaurs, than they do living crocodiles. Also, living crocodiles show a number of the remarkable adaptations related to the advanced metabolisms seen in modern birds, including four-chambered hearts and highly efficient, birdlike lungs. To add to this, there is even molecular evidence that many of the genes which are responsible for feathers in living birds exist in crocodilians. So it is at least plausible that feathers may extend all the way back down the family tree to the base of Archosauria, and that they were subsequently lost when the group that evolved into living crocodiles acquired their primarily aquatic habits.

STEGOSAURUS STENOPS

ANOTHER OF THE BIG FIVE THAT ROLLS OFF THE TONGUE OF EVERY 10-YEAR-OLD. *STEGOSAURUS*, THE PLATED LIZARD, BELONGS TO A GROUP OF PREPOSTEROUS-LOOKING ANIMALS. MOST HAD SMALL HEADS, BIG BODIES, ALL SORTS OF BONEY PLATES, AND SPIKES.

Stegosaurus was first discovered in the Late Jurassic rocks of the Morrison Formation near the town of Morrison, Colorado, USA. It was formally named in 1877 by O.C. Marsh. Since then, its remains have been found in most other Morrison Formation localities, including coeval rocks in Portugal. It was a large animal, up to 9m (29½ft) in length, and up to 7 tonnes in weight. *Stegosaurus* and its relatives had ridiculously small teeth, and many had throats containing small bones called osteoderms which formed a sort of protective chainmail in this vulnerable area.

Two things have fascinated the public since the discovery of *Stegosaurus* – its plates and its spikes – and these have also formed the basis of a great deal of research. Arguments for the function of the plates range from defensive armour to (incredibly) flaps to propel the animal off the ground. The function of the tail spikes (called thagomizers) is similarly debated. When these animals were first discovered, the arrangement of the plates was not apparent. Early on it was thought that they lay on the sides of the body in a protective, shingle-like arrangement. Later, it was suggested that they formed a single row of plates along the animal's back. A paired double row of plates

soon followed, before today's arrangement of alternating plates on the midline was settled upon.

As mentioned, some unusual ideas have been put forward regarding the function of the plates and spikes. Beyond the obvious defence and display arguments, a novel hypothesis is that they acted as thermoregulators. Citing the large amount of vascularization on the surface of the bone, researchers postulated that they acted like a car radiator and helped the animal dissipate excess heat. This is what happens with the large ears of some animals found in hot regions, such as desert rabbit species and African elephants. While this sounds plausible, it fails the comparative test.

In the time since the first *Stegosaurus* specimens were found in the American West, a number of closely related animals have been found. They all have virtually the same body plan, except that the shapes, sizes and architecture of the spikes and plates differ immensely. Some stegosaurs, like *Huayangosaurus*, have large spikes on their shoulders. This developed to an

> A fanciful depiction of *Stegosaurus*. Pterosaurs are grooming it, removing parasites.

↳ The tail end of *Stegosaurus* in a modern pose — the tail held high and the plates imbricate.

∨ *Stegosaurus* is among the largest and best known of stegosaurs. This specimen was discovered by Museum paleontologists in 1901 and remains on display today.

almost absurd degree in the Chinese Late Jurassic stegosaur *Gigantspinosaurus*, where the shoulder spike extends almost the entire length of the torso. Others, like *Dacentrurus*, have plates that gradually grade into thagomizers at the base of the tail; it also has very small back plates that would have provided very little in the way of a radiating surface. Other stegosaurs have primarily spikes and no plates. Consequently, the notion that these evolved for thermoregulation is spurious. So why were they there? While the thagomizers may have had some defensive capability, the rich diversity of spikes and plates really points to a display function.

A final word about *Stegosaurus* is that it is often referred to as the dinosaur with two brains. When the animal was first described, it was noted that there was an expanded opening in the pelvis over 20 times as large as the actual brain cavity in the animal's skull. At the time, the theory was that this may be a "second brain" which helped control the rear end of the animal. It turns out that this structure is also found in many other dinosaurs, including sauropods. Currently, it is thought that it housed a glycogen body that is found in living birds. The function of this body, especially in relation to the spinal column, is unknown.

> An early reconstruction of *Stegosaurus* with the plates in a line and the tail held down.

SAUROPELTA EDWARDSORUM

EARLY CRETACEOUS
CLOVERLY FORMATION
WESTERN NORTH AMERICA

SAUROPELTA BELONGS TO A GROUP CALLED THE NODOSAURIDAE, WHICH IS THE CLOSEST RELATIVE OF THE ANKYLOSAURS. AS RARE AS ANKYLOSAURS ARE, THEIR MORE PRIMITIVE RELATIVES, THE NODOSAURS, ARE EVEN LESS COMMON. NONE OF THESE ANIMALS ARE THOUGHT TO HAVE BEEN THAT NUMEROUS BECAUSE THEIR REMAINS ARE RARE WHEREVER THEIR FOSSILS ARE FOUND.

They are very similar to ankylosaurs except that they lack a tail club and their armour is more like chain mail. Sometimes the tail can be very long, as in the case of *Sauropelta*, where it is almost half the length of the body. Nodosaurs are found throughout Laurasia, with one outlier in Antarctica; most of the best specimens have been collected in Western North America.

Nodosaurs are thought to have had similar habits to ankylosaurs. Until recently, the best known nodosaur specimen was collected by Barnum Brown in central Montana, and was named *Sauropelta*. Unlike most ankylosaurs, some nodosaurs had extraordinarily large spikes emanating from their shoulders – a sort of idealized gladiatorial look.

A fantastic dinosaur specimen, one of the best finds in recent years, was discovered accidently in an open pit asphalt mine in Alberta, Canada. Observant heavy equipment operators noticed that their excavator had crashed into fossil bones and they notified paleontologists who rushed out to the site. Vertebrate fossils are not unknown or uncommon in sediments in this area, but almost all the animals that have been found have been marine reptiles such as plesiosaurs, which are not dinosaurs. What awaited the scientists was a surprise: instead of a marine animal, they were presented with the remains of a dinosaur, and a very well-preserved one at that, with significant soft tissue preservation, especially the skin. The animal was named *Borealopelta*, and it is by far the best nodosaur found thus far. The skin and plate arrangements clearly show that the armour was not a hard shell – instead it was a flexible mail that could take blows or bites without cracking, and this flexing absorbed and dissipated much of the energy that a predator's weapons might inflict on the animal.

> While not *Sauropelta*, this amazing specimen of the closely related *Borealopelta* was discovered in 2011, and officially named in 2017. It is by far the finest specimen of a nodosaur yet discovered.

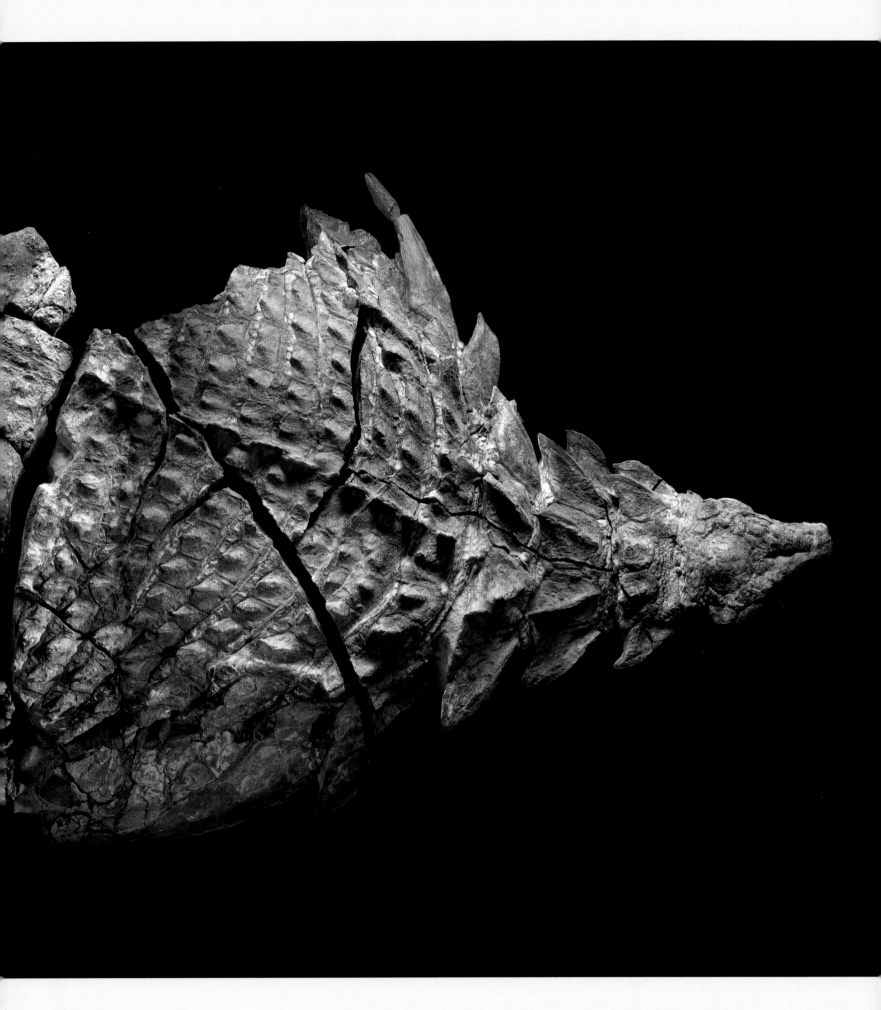

So how would a terrestrial dinosaur have found itself deposited on the sea floor and become so well preserved? In order for this to occur, a strict set of conditions had to take place: the nodosaur would have had to die on land, probably next to a river, which would then have carried the carcass out to sea. As the animal began to decompose, the process would have produced a great deal of gas, causing the animal to float. Eventually the body would have ruptured, and the gas would have escaped, causing the body – weighed down by its heavy, boney armour – to sink to the sea floor. Paleontologists have dubbed this process "bloat and float".

Another such "bloat and float" specimen is a member of the closely related ankylosaur group. Found in marine sediments near San Diego, California, USA, this specimen was exposed on the sea or lagoon floor long enough for barnacles to begin growing on its plates.

The conditions on the sea floor need to be just right for an animal to be preserved and form a fossil. First, the water cannot be too deep. The hard component of bones is the mineral calcium phosphate. At the Earth's surface calcium phosphate is very stable; however, if bones find themselves in deep ocean basins (anything over about 2½miles/4km) they will start to dissolve. Secondly, the water cannot be too shallow, because in shallow water many organisms will actively feed on bones and other aspects of decaying carcasses. Under perfect conditions, after the bones sink into water that is not too deep, they need to become quickly buried to allow for the preservation of a remarkable specimen.

∧ Barnum Brown, probably the greatest dinosaur hunter in history, in the field in Montana in the early twentieth century.

> *Sauropelta* being worked on in the Museum during a move of the dinosaur collections.

EUOPLOCEPHALUS TUTUS

LATE CRETACEOUS
DINOSAUR PARK FORMATION
WESTERN NORTH AMERICA

ANKYLOSAURS WERE NEVER VERY DIVERSE, AND THEY ALL LOOKED VERY SIMILAR EXCEPT FOR THEIR SIZE, BUT AMONG THEM THERE ARE SOME REALLY REMARKABLE SPECIALIZATIONS AND *EUOPLOCEPHALUS* IS A GOOD EXAMPLE.

It is closely related to and looks much like *Ankylosaurus*. It was slightly smaller (7m/23ft), and was the same size as many other Late Cretaceous ankylosaurs (other than *Ankylosaurus*, see p.178) from Asia, such as *Tarchia*, *Minotaurasaurus* and *Pinacosaurus*.

Euoplocephalus is not as heavily armoured as *Ankylosaurus*, its skull is not as triangular in shape, and it lacks the large horns at the rear corners of the head. The skull measured 41cm (16in) in length and 40cm (15¾in) in width, proportionally different to *Ankylosaurus*. In contrast to the fairly close relative *Pinacosaurus*, the remains of most other ankylosaurs, like *Euoplocephalus* and other Asian varieties, have usually been found as isolated and disarticulated specimens. Conversely *Pinacosaurus* specimens have been found as associated individuals at several localities in Central Asia. This is particularly true of juveniles, which are often preserved as flocks in catastrophic death assemblages. Curiously, when this happens, all of these single-age cohorts are found oriented in the same direction parallel to one another.

There are two defining characteristics of ankylosaurs beyond their armour: one well-known and one esoteric. It is not commonly known that ankylosaurs have extraordinarily unusual nasal passages. In essence, after air entered the nose it would take a labyrinthine route through the head before it entered the pharynx. Why these animals had such a complex nasal cavity is not understood; if it was to give them an enhanced sense of smell, we would expect that this would be associated with enlarged nasal lobes in the brain, but as CT analyses show, the nasal sensory areas of the ankylosaur's brain are typical of other dinosaurs. Another suggestion is that they were used for a mammalian type of

∧ A skeleton of *Euoplocephalus*. This rather grotesque-looking dinosaur was an uncommon herbivore in the Late Cretaceous of western North America.

< *Euoplocephalus* was first discovered in the badlands of Alberta, Canada, as is true of many other Late Cretaceous dinosaurs featured in this book – *Albertosaurus*, *Styracosaurus*, *Hypacrosaurus*, *Saurolophus*, *Corythosaurus*.

water and heat balance. On cold or extremely humid days you can see your breath as a cloud of steam – warm, moist air is passing out of your nose through a system of bones called turbinals, which help you thermoregulate. The complicated architecture of ankylosaur nasal passages could have provided a similar function. A question to ask here is: other dinosaurs were thought to have the same sort of physiology as these animals, so why didn't they have them? Finally, the nasal passages have been considered as noise-producing organs. While some research has been carried out to suggest that the ears were adapted to hear low-frequency sound, there is no comparative test across different dinosaur taxa that can be made at this time.

The better-known ankylosaur feature is their tail clubs: all of these animals wielded mighty mace- or bludgeon-like structures on the ends of their tails. Some, as in *Euoplocephalus* and *Ankylosaurus*, were dense, tripartite bone clubs held aloft via a very inflexible tail that was supported by thick tendons and fused vertebrae. These support structures were restricted to the back section of the tail, so it is likely there would have been considerable lateral movement if the tail was swung in a defensive fashion. But while the possibility of a mighty blow exists, no rigorous biomechanical modelling of the tails of these animals has yet been completed, and certainly there is much variation in tail size and shape among the different species. *Pinacosaurus*, for instance, has a tail more like a mace than a bludgeon, so there could easily be a high element of display function, either for ritualized interspecies combat, species recognition, or both.

ANKYLOSAURUS MAGNIVENTRIS

LATE CRETACEOUS
HELL CREEK FORMATION
WESTERN NORTH AMERICA

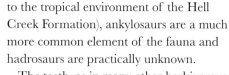

ANKYLOSAURUS IS ONE OF THE LARGEST ARMOURED DINOSAURS, MEASURING 8M (26¼FT). IT IS A VERY RARE ANIMAL AND ONLY A FEW SPECIMENS ARE KNOWN, BUT WHAT IS KNOWN IS PRETTY IMPRESSIVE.

The skulls were enormous, up to 65cm (25½in) long and 75cm (29½in) wide, and when viewed from above were fairly triangular in shape. The nostrils flare out to the sides and the corners at the back of the skull sport large horns. At the end of the upper jaw there was a beak rather than teeth, and it is presumed that the end of the lower jaw was also edentulous (as in *Pinacosaurus*), but this element is not preserved in any of the known specimens. Like many thyreophoran herbivores in the Hell Creek Formation, except for hadrosaurs and ceratopsians, these animals were relatively uncommon when they were alive. It is interesting that in other areas, like the Gobi Desert, which was thought to be arid in the past (as opposed to the tropical environment of the Hell Creek Formation), ankylosaurs are a much more common element of the fauna and hadrosaurs are practically unknown.

The teeth, as in many other herbivorous dinosaurs of the time – such as pachycephalosaurs, and nodosaurs, and stegosaurs before that – were preposterously small. This makes what they ate something of an enigma: how could an animal this size sustain itself and process so much food with such feeble teeth? If it had hindgut fermentation (as in Galápagos tortoises today), some have suggested that *Ankylosaurus* could have survived on about 70kg (154lb) of plants and fruit each day, which is still an impressive amount. One apparent feeding adaptation is the very large tongue bones that have been found in the close *Ankylosaurus* relative *Pinacosaurus*. These tongue bones are thought to have supported a thick, bulbous tongue; today some animals, such as salamanders, although much smaller, have bulbous tongues and use them to pick

up food off the ground. The sideways-facing nostrils of *Ankylosaurus* may also suggest that the animal had its nose to the ground while acquiring food.

Like all of its relatives, *Ankylosaurus* was a highly armoured animal. Unlike the so-called "armour" present in many other dinosaurs, it is more than reasonable to say that this armour provided a great deal of protection. Its head was covered with thick, bony plates called osteoderms, which in the adults were fully fused to the skull. Behind the skull – although incompletely known from *Ankylosaurus*, but well

known from *Pinacosaurus* – there were bony crescents that protected the neck. The plates on the body numbered more than 100; these were relatively thin and ranged in size from 1cm (⅜in) to more than 35cm (13¾in). They did not fuse to any bones and did not imbricate, but instead lay completely within the skin, which would have provided heavy, yet flexible protection. *Ankylosaurus* also possessed a large tail club like its relatives (see *Euoplocephalus*, p.176). With its low, quadrupedal gait it must have looked like a formidable, slow-moving, vegetarian tank as it lumbered across the land.

∧ A reconstruction of *Ankylosaurus*. It is difficult to imagine that an animal of this mass could sustain itself considering its tiny teeth and small mouth.

⌐ The large and formidable tail club of *Ankylosaurus*. Its function has not been determined conclusively. All we know is that it was held aloft.

< Anklyosaur armour in the field, just collected by Barnum Brown during the Red Deer River excavations in Alberta, Canada.

PACHYCEPHALOSAURUS WYOMINGENSIS

LATE CRETACEOUS
HELL CREEK FORMATION
WESTERN NORTH AMERICA

ONE OF THE MOST UNUSUAL OF THE ORNITHISCHIAN DINOSAURS IS *PACHYCEPHALOSAURUS*. THE NAME MEANS "BONE-HEADED LIZARD" AND ONE LOOK AT THE SKULL GIVES YOU A GOOD IDEA WHY: IT LOOKS LIKE AN ANIMAL WITH HALF A BOWLING BALL ON ITS HEAD.

This big head was made of solid bone over 25cm (10in) thick surrounded by a crown of large spikes that even extended onto the nose of the animal. Inside this dense mass of bone was a small brain. *Pachycephalosaurus* was a bipedal animal, and is a primitive member of the group Marginocephalia, which also includes horned dinosaurs. These dinosaurs had small, laterally compressed teeth, and were incapable of eating strong, fibrous plants.

Pachycephalosaurus is the largest pachycephalosaur and was about 3m (10ft) long. It, like nearly every other pachycephalosaur, is known from fairly fragmentary remains and no complete specimen of any member of this group of dinosaurs has yet been recovered. Usually what is found of these animals is either their distinctive teeth, or the very hard, dense skull cap, which, because it is so hard, is more likely to be fossilized.

Never very diverse, remains of these animals have been found primarily in North America and Asia. The paucity of specimens suggests that pachycephalosaurs were never very numerous during their lifetime.

Since the first discovery, the function of the bony dome on the skull of *Pachycephalosaurus* and its relatives has been a contentious topic. One common idea is that the animals used them as battering rams to defend territories or attract mates, behaviour that is similar to that of bighorn sheep today. However, there is evidence that this was not the case.

First, it has generally been considered that not all pachycephalosaurs had domed heads. While some, such as *Pachycephalosaurus*, *Prenocephale*, *Stegoceras*, *Tylocephale* and others certainly did, several flat-headed species such as *Homalocephale* have also been identified. However, new evidence seems to suggest that these flat-headed forms may in fact be juveniles of the dome-headed varieties. These specimens show unmistakable signs of skeletal immaturity, as the sutures delimiting the individual bones are still visible, and many of the holes (fenestrae) on the back of the skull are still apparent. A characteristic of pachycephalosaurs is that these holes become covered by the thick dome as the animal grows to maturity.

The theory that the animals did butt heads is supported by the observation that the bones of the neck are reinforced at their articulations to make the neck rigid, yet the neck still retained its S-shape. Moreover, a significant number of dome-headed pachycephalosaur specimens possess lesions. This type of lesion also occurs in living animals and is the result of severe trauma to bone surfaces. None of the flat-headed pachycephalosaurs showed this type of injury, which may suggest that if head-butting occurred, it was restricted to mature or male members of these species.

While spectacular, pachycephalosaurs remain a very enigmatic group. Until additional material is found, much remains unknown about the biology of these curious animals.

∧ *Pachycephalosaurus* and its relatives are some of the weirdest dinosaurs with their domed skulls and grotesque appearance.

↖ The type specimen of *Pachycephalosaurus* displaying the high-domed skull that is a characteristic of some members of the group.

> Shown here in longitudinal section, it is apparent that unlike many other dinosaurs which have air-filled crests on their skulls, the skull of *Pachycephalosaurus* is solid bone (top) above its small brain (bottom).

< Although the skull of *Pachycephalosaurus* is well understood, the rest of its skeleton remains poorly know.

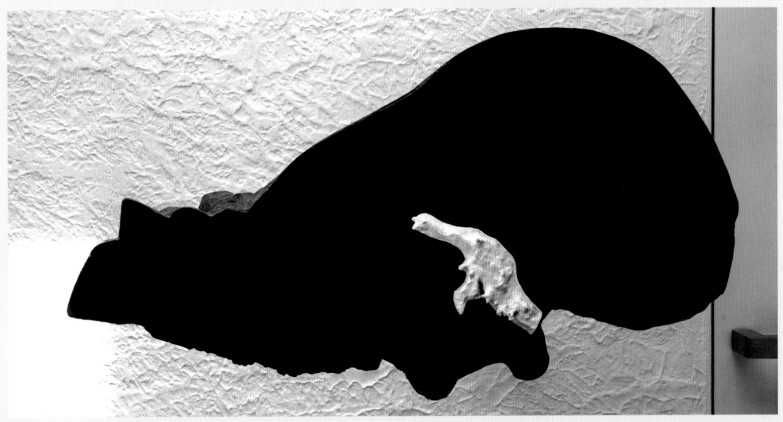

PSITTACOSAURUS MONGOLIENSIS

EARLY CRETACEOUS
ASHILE FORMATION
MONGOLIA

PSITTACOSAURUS IS A PECULIAR-LOOKING DINOSAUR. ITS NAME MEANS "PARROT REPTILE", WHICH REFERS TO ITS HIGHLY UNUSUAL SKULL FEATURING A PARROT-LIKE BEAK. IT WAS A SMALL ANIMAL, ONLY ABOUT 1M (3¼FT) IN LENGTH, AND LIVED ABOUT 125 MILLION YEARS AGO IN WHAT IS NOW EAST AND EASTERN CENTRAL ASIA. THERE IS ALSO A REPORT BASED ON VERY FRAGMENTARY EVIDENCE OF AN INDOCHINESE OCCURRENCE.

The first specimens were found by the American Museum of Natural History's Central Asiatic Expeditions in 1922 and described a year later by Henry Fairfield Osborn. It was named *Psittacosaurus mongoliensis*, after the country of Mongolia in which it was found. Unlike so many other dinosaurs, the genealogical placement of this animal was immediately recognized. Several features of the skeleton indicated early on that it was a primitive member of the group that includes the advanced ceratopsians such as *Triceratops* and more primitive taxa like *Protoceratops*.

Since its original discovery, literally thousands of specimens have been found and several species of *Psittacosaurus* have been named, primarily in the last decade. Most of these are from China, although significant new remains have also been found in Mongolia and Siberia. The majority of Chinese specimens have come from the Jehol Biota, a series of sediments in Northeastern China, which has preserved so many of the great specimens that have revolutionized dinosaur paleontology in the last 20 years. One specimen is particularly important, and today it is housed at the Senckenberg Museum in Frankfurt, Germany.

From a scientific perspective it is extraordinarily interesting, but ethically the find is highly questionable: it was smuggled out of China in contravention of Chinese laws restricting the commercial fossil trade. Each year the world's largest fossil trade show takes place during the last week of January in Tucson, Arizona, USA. While the show is now much more carefully regulated, in the early 2000s it was a bizarre mix of the legitimate, the fake and the illegal. While some illegal trading still takes place (indeed, there were arrests

7 A nearly complete specimen of *Psittacosaurus*. The animal gets its name (meaning "parrot dinosaur") from its avian-like beak.

at the 2016 show for the illegal smuggling of Chinese dinosaur eggs), most of it is above board. This was not the case previously, however, and Tucson is where a fantastic *Psittacosaurus* specimen found a buyer, and it is this specimen that found its way to the Senckenberg. While representatives of the Chinese government have repeatedly called for its return, their efforts have so far been unsuccessful. What makes this specimen so special is that it preserves soft tissue. While many of the Liaoning specimens preserve such structures, the Senckenberg specimen is the first ornithischian dinosaur to show definitive evidence of filamentous structures.

These structures consist of long bristles that form a crest or comb on the animal's tail, much like the crest on the tail of the extant giant anteater. When these were first described, it was not clear what these structures were – it was even suggested that they were the remains of a plant that had fallen on top of the skeleton prior to fossilization. However, because of the many other discoveries of genuine soft fossil tissue in the Jehol beds, it is now firmly established that these are actual structures, and that they are homologous (meaning they have the same evolutionary origin) to the feathers of modern birds and the filamentous body coverings of so many other dinosaur specimens.

∧ *Psittacosaurus* was a very common animal in its day, and low on the food chain. It was preyed on by primitive mammals such as *Repenomamus*. *Psittacosaurus* remains have been found in the fossilized gut contents of this predator.

> *Psittacosaurus* was a small dinosaur, only about a metre long. It was probably primarily bipedal.

In rare cases, catastrophic events kill animals and preserve them simultaneously. This scenario has been discussed for the small troodontid *Mei long* (see p.117) and many of the specimens collected at the Ukhaa Tolgod locality. North of Beijing there is a series of rocks called the Lujiatun beds. Although not universally accepted, many geologists consider these rocks to be the result of a catastrophic volcanic eruption, which rained volcanic ash and created low-temperature pyroclastic flows that entombed many animals while they were still alive. *Psittacosaurus* were among them and are probably the most common animals found at this site. Many of these specimens were discovered in classic sleeping positions – crescent-shaped and with their heads neatly tucked into their bodies. As with *Mei long*, this suggests that the animals may have succumbed to toxic gases produced by volcanic vents before their burial.

Other specimens from the same beds show accumulations of several individuals. These are uniformly juvenile and sometimes occur in large numbers; the largest known association reported to date is 34. While all of these were the same age (based on size), accumulations of mixed size have also been collected, suggesting that these animals were gregarious in nature, especially when young. The presumption is that these accumulations occurred when all of the members of the group perished at once in a burrow collapse or were buried following a volcanic event as

described above, and this provides powerful evidence of sociality.

Whenever vertebrate fossils are as well-known as *Psittacosaurus* is, there are bound to be some surprises. One of these is proof that *Psittacosaurus*, a herbivore at the lower end of the food chain, was the prey for carnivorous animals. Multiple cases from around the world have demonstrated that dinosaurs, particularly young ones, were preyed on by other dinosaurs and crocodilians. The fossil beds of Northern China have also preserved many fossil vertebrates aside from dinosaurs, including several species of fossil mammals that were contemporaneous with the dinosaurs. However, during the Mesozoic era, they were not very speciose and were never much larger than a domestic cat. It was only after the disappearance of non-avian dinosaurs that they proliferated into ecological niches previously occupied by dinosaurs and sea-going reptiles. One of the largest Mesozoic mammals, known as *Repenomamus*, lived about 124 million years ago in what now is modern China. It is a member of the primitive mammalian group Triconodonta, which has no living relatives. It was unusually large for a Mesozoic mammal, weighing up to 14kg (31lb) and growing to about 1m (3¼ft) long. In one extraordinary *Repenomamus* specimen, the remains of a small *Psittacosaurus* was found. The chick was semi-articulated, suggesting that *Repenomamus* swallowed it in big chunks with little chewing. This provides the first incontrovertible evidence that mammals occasionally preyed on dinosaurs, which were the dominant form of terrestrial life at that time.

PROTOCERATOPS ANDREWSI

LATE CRETACEOUS
DJADOKHTA FORMATION
MONGOLIA

FOSSILS OF DINOSAURS ARE VERY RARE. OF THE LESS THAN 1,000 NAMED, NON-AVIAN DINOSAURS, ABOUT ONE-THIRD TO A HALF ARE KNOWN FROM SINGLE INDIVIDUALS, AND POOLED TOGETHER, THE GREAT MAJORITY OF DINOSAUR FOSSILS ARE KNOWN FROM FEW AND FRAGMENTARY REMAINS.

This greatly hinders our ability to understand the biology of these animals, such as the physical changes that took place during growth, how fast they grew, or whether characteristics that may indicate differences between the sexes can be determined. Conversely, for *Protoceratops* there is an abundance of riches, as thousands of specimens ranging from single teeth to complete skeletons have been discovered, all of them from the Gobi Desert in Mongolia and north-central China (Inner Mongolia).

The first specimens were collected by the Central Asiatic Expeditions of the American Museum of Natural History in 1922. These were found at the legendary Flaming Cliffs locality. Called Bayn Dzak in Mongolian, the name means "rich in dzak" – dzak being the small trees that form a forest at the base of the cliffs. The name *Protoceratops* means "first horned face", but at the time of the discovery, Museum paleontologists had no way to accurately date the age of the sediments at the Flaming Cliffs site. *Protoceratops* is a very primitive horned dinosaur, part of the group that is related to, but not a member of, the classic group of Late Cretaceous horned dinosaurs of the North American West. These

animals, including *Triceratops*, *Chasmosaurus*, *Monoclonius* and *Styracosaurus*, were much bigger (some were up to 8m/26¼ft) than *Protoceratops*, which fully grown was the size of a large pig. The paleontologists of the Central Asiatic Expeditions were very familiar with advanced horned dinosaurs because they had excavated many of these species in the USA during the early part of the twentieth century.

The contemporary thinking was that because *Protoceratops* was one of the most primitive members of the horned dinosaur group known at the time, it must have predated the more advanced members. Therefore, they conjectured that the Bayn Dzak rocks were a lot older than sediments in the USA that entombed their more advanced relatives. It turns out that they were wrong, as using "degree of primitiveness" is a very poor way to determine when a species lived. For example, let's imagine that two groups of aliens visit Earth and sample the mammal fauna in

⌐ A juvenile specimen (only 20cm/8in long) of *Protoceratops*. This small animal was buried alive, accounting for its remarkable preservation.

two places – the Amazon and Madagascar. The aliens might reasonably conclude that the animals in Madagascar were the ancestors of, or at least existed earlier in time than the super-specialized fauna of the Amazon, because the Madagascan species look much like the animals that lived 50 million years ago. They, like *Protoceratops*, are very primitive – almost like living fossils. But evolution and the history of Earth do not work like this – evolution is not linear, it is a bush-like genealogy – and that is why we can have primitive animals like the egg-laying mammals echidna and platypus existing alongside rats, cows, and humans.

Protoceratops is also notable for what it may have inspired. Adrienne Mayor is a renowned folklorist and a research scholar at Stanford University. Although controversial, her conjecture is that *Protoceratops* may have been the inspiration for the griffin in classical western mythology. While the resemblance to Ancient Greek rhyton drinking cups and gargoyles is remarkable, there is a backstory.

Proximate to the area of Southern Mongolia where *Protoceratops* is found, there is a famous mountain called Altan Ula, which means "gold mountain" in Mongolian. For thousands of years this area has been a known source of gold, and ancient mining is both documented and apparent. It is only recently that these early prospects have become one of the largest gold and copper deposits in the world. In mythology the griffins were the protectors of treasure. Is this a coincidence? Probably, but it is an interesting thought.

The plethora of specimens allows us to do a lot of science. For instance, we know that living birds spend less time in the egg during brooding than reptiles of the same size. A 100g (3½oz) lizard develops in the egg for 140 days, while a bird of the same size takes 25 days. How can we determine how long this would have taken in a non-avian dinosaur? Luckily, we have found a specimen. In 1997 we excavated a nest of eggs at Ukhaa Tolgod in Mongolia. What was remarkable about these

∧ The Central Asiatic Expedition camp at Oshii in Mongolia. It was here that many important Early Cretaceous specimens were collected.

⌐ A remarkable specimen of *Protoceratops* from Bayn Dzak. This nearly mature animal was probably buried alive and died curled up in a seemingly defensive posture.

> The skull of a large adult *Protoceratops* in situ at Bayn Dzak. The Central Asiatic Expeditions made extensive collections of these animals. These fossils compose a growth series from hatchlings to senescent adults.

eggs is that each one contained an embryo, and using new technology we have been able to calculate how many days these embryos had been in the egg before their demise.

Toothed amniotes (the group that includes mammals, reptiles and birds that lay their eggs on land) all build teeth in the same way. Usually teeth begin to develop between 40 and 60 per cent of the way through the total incubation period. There are physical constraints on the speed and the way in which teeth are formed. The majority of a tooth is composed of a relatively soft material called dentine which is cloaked in a hard coating of shiny enamel. The mechanics of tooth formation only allow about 20 microns of tooth dentine to be built per day – about 0.0003cm. In addition, each day a line is laid down – analogous to a tree ring – although the mechanism for this is not very well understood.

The Ukhaa Tolgod embryos have allowed us to study this and we were able to use a combination of CT scanning (a three-dimensional X-ray) and optical microscopy to count the lines. Our calculations show that

the young *Protoceratops* spent about 80 days in the egg. Considering the size of the egg, if the embryo was a bird we would predict that it would have been in the egg for 40 days. If it was a non-dinosaurian reptile like a lizard, we would expect it to be about 150 days. What we found was that it was in between – unsurprising considering the fact that *Protoceratops* is more closely related to modern birds than it is to any non-dinosaurian reptiles.

The number of *Protoceratops* specimens has also allowed us to understand a great deal about growth. This was noted early on, and one of the most studied and startling exhibits on display at the Museum is a growth series of *Protoceratops* skulls amassed during the Central Asiatic Expeditions. It is a linear array of a dozen specimens that range from a 10cm (4in) long juvenile with a small frill, to an animal with a skull length of 100cm (40in). Later expeditions have added another 100 or so specimens which range from the embryos described above to larger animals. Using new mathematical techniques, analyses of these specimens are in the pipeline, and we may be able to determine whether males and females were different, and just how variable these animals were in life.

> *Protoceratops* was a pig-sized quadrupedal herbivore that shared some features with advanced horned dinosaurs, such as the boney frill and enlarged cheekbones, but lacked the large horns of those species.

TRICERATOPS HORRIDUS

LATE CRETACEOUS
HELL CREEK FORMATION AND OTHERS
NORTH AMERICA

AMONG THE BIG FIVE OR SO MOST FAMILIAR DINOSAURS, *TRICERATOPS* IS NEAR THE TOP IN TERMS OF NAME RECOGNITION. IT IS A STRANGE-LOOKING ANIMAL, AND BECAUSE IT IS VERY COMMON AND APPEARS IN MANY MUSEUM EXHIBITIONS WORLDWIDE, THERE HAS BEEN EXTENSIVE RESEARCH INTO THIS TAXON. BUT FIRST A FEW SPECIFICS.

Triceratops was a large (up to 9m (29½ft in length), efficient herbivore that lived at the end of the Cretaceous period and was one of the dinosaurs that would have witnessed the after-effects of the asteroid hitting the planet at the terminal Cretaceous event. Its proportions are massive, and a single skull can be over 2.5m (8¼ft) in length. As with other ceratopsian species, it is gargoyle-like in appearance.

As the name implies (it translates as "three-horned face"), the head features three large horns – one above each eye and the other at the tip of its nose. The first discovery was made by a school teacher in the vicinity of Denver, Colorado, USA, in 1887, who found some very large fossil horns. These horns were sent to O.C. Marsh at Yale University, and on preliminary examination, Marsh believed they represented the horns of a giant extinct bison. However, after more material was collected, Marsh recognized that instead of a mammal, these were the remains of a large horned dinosaur. Much taxonomic confusion followed, but the name *Triceratops* was finally decided upon.

Subsequently, a great deal has been learned about this animal and there has been controversy surrounding some of this research. What is known for certain is that these animals existed in great numbers and, together with some hadrosaurs, they were the most prevalent herbivores of their age. Literally hundreds of these animals have been found, and they are so common that they are not always excavated. We also know quite a bit about how fast they grew, but we do not know exactly how long it took them to reach maturity. Even though the food they ate was of higher quality than what was available to early dinosaurs (high-calorie angiosperms had developed by this time), they must still have consumed enormous amounts. Their jaws contained hundreds of teeth, annealed into dental batteries that formed a single shearing surface. In contrast to the sauropods, these animals processed their food orally. And unlike many other ceratopsian dinosaurs, there are no *Triceratops* bone beds, which suggests they were solitary animals when alive.

Triceratops teeth were extremely advanced. These animals were eating huge amounts of food and had to replace their teeth quickly. Animals today, such as horses and rodents, which eat coarse vegetal food and pick up a significant amount of dirt while doing so, wear their teeth down at a fast rate. The evolutionary solution to this in mammals is to have high crowned, rapidly growing teeth. Dinosaurs had lots of teeth that replaced themselves. Some ornithischian dinosaurs took this to the next level. In our teeth we have four tissue types – the familiar ones such as dentine, enamel, cementum and pulp. Ceratopsians, and also advanced hadrosaurs, had elaborations on this totalling seven types of dental tissue, making them the most complex teeth in the tetrapod world. These were highly evolved specialists, consuming fodder on a large scale, and they should in no way be viewed as primitive reptiles.

Triceratops is named because of its horns, and together with its awesome frill, these have led to a great deal of science speculation and myth. When children first encounter a model or drawing of *Triceratops*, they interpret it in a similar way to many in the research community: it was an aggressive, three-pronged, rhinoceros creature with a shield covering its neck. This is uncertain. *Triceratops* would certainly have been more fearsome in life than death, as its horns would have been appreciably larger. Some of the fossil horns from above the eye are over 1m (3¼ft) long. In life, these would have been up to one-third longer as they were most likely covered with a keratinous sheath. (On cows today, the horn we see is much larger than the bony core supporting it.)

Nevertheless, while function is something that is very hard to judge when you can't

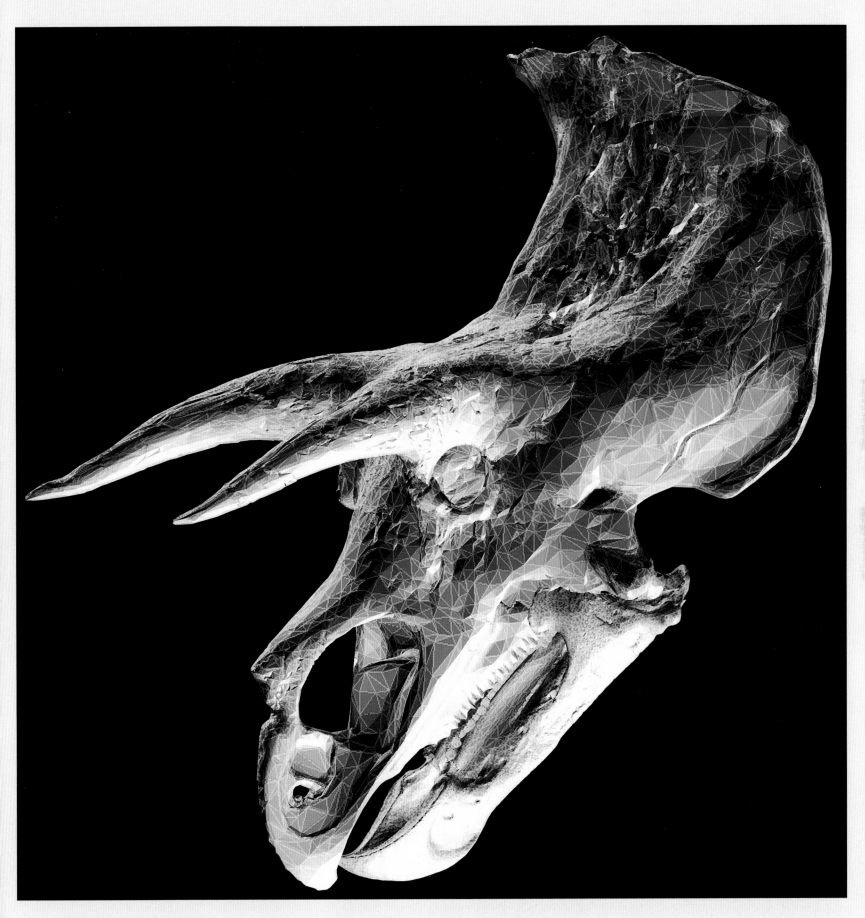

∧ *Triceratops* is the best
known of the horned dinosaurs.

immediately observe the animal, we can throw out a few facts. We know that *Triceratops* was a prey target of *Tyrannosaurus rex* from several known specimens that have extensive injuries sustained as a result of predation by large theropods. Some of these may be the marks of a scavenger, but others clearly are not as there are signs of healing and scar tissue around the wounds.

Regarding the horns, there are two theories – echoed in other chapters in this book – display or defence. In the current view, display wins. Even if a *Tyrannosaurus rex* was foolish enough to engage an adult *Triceratops*, the *Triceratops* would probably have lost. Its bones would have exploded, and the few specimens that we have of *Triceratops* with bite mark injuries that have healed, were lucky individuals indeed. The head and frill structures, like elaborate cranial ornaments in animals today, were much more likely to have been used for flash and interspecific combat than for fighting off predators.

An idea proposed in 2010 was that *Triceratops*, together with another contemporary ceratopsian *Torosaurus*, represented the same species. *Torosaurus* differs from *Triceratops* in the fact that the large frill that exists on the back of the skull is perforated by large fenestrae (meaning "windows" in Latin), which are large excavations or holes in the bones. In many ceratopsian species that have frill fenestrae (*Protoceratops*, for example), these do not appear in the youngest specimens, and the skulls become perforate with age. While this is intriguing, when you consider that the largest *Triceratops* skulls that exhibit no fenestrae are about the same size as the fenestrated *Torosaurus* skulls, this idea runs into problems. Most dinosaur researchers still consider these to be very different species.

< `Triceratops` had lots of teeth that were organized into a dental battery to orally process food.

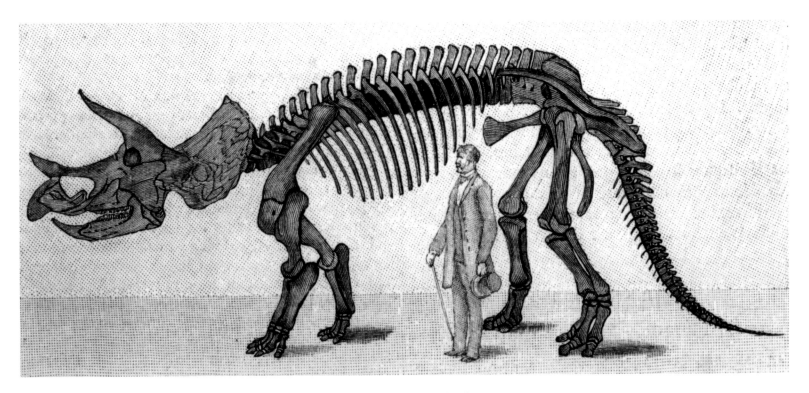

∧ A historic rendering of the *Triceratops* skeleton.

∨ Charles R. Knight's mural of a Late Cretaceous landscape. The posture of the animals, their anatomy and their environment are all dated.

MONOCLONIUS NASICORNUS

LATE CRETACEOUS
WESTERN NORTH AMERICA
JUDITH AND DINOSAUR PARK FORMATIONS

JUST THE NAME *MONOCLONIUS* (MEANING "ONE-HORNED FACE") CAUSES ANGST FOR MANY DINOSAUR PALEONTOLOGISTS. EVERYTHING ABOUT IT SPEAKS TO CONFUSION, AND EVEN THE WIKIPEDIA ENTRY REFERS TO IT AS A "DUBIOUS GENUS OF HERBIVOROUS CERATOPSIAN DINOSAUR".

First the specifics. It was originally named by E.D. Cope in the early days of North American dinosaur collecting. When it was collected, the area of the type locality was not the safest place to be – the Indian Wars were in full swing and Custer was killed less than 100 miles (160km) away that same year. Nevertheless, many more specimens were collected. Because *Monoclonius* was one of the first ceratopsian dinosaurs known, a number of other ceratopsian specimens were assigned to this taxon. One of the best specimens can be found at the American Museum of Natural History; it contains nearly every bone in the body, including the delicate bones inside the eye, which supported the lens.

But herein lies the problem: so many specimens have been ascribed to *Monoclonius* that it is difficult to refer any single specimen to this genus. Something else that complicates matters is that a very similar animal named *Centrosaurus* is found in the same beds. *Centrosaurus* has a very similar skull shape, and the adornment on the skull is also similar but is expressed to a different degree. Compounding the issue is the fact that these are very common animals. The more common the animal, the greater the sample size, and the more variation is introduced. Using a human analogy, if we were to compare the fossils of a professional basketball player with a jockey, or a sumo wrestler with a ballerina, we might very well think that the bones were from different species. But because there is a complete range of sizes and shapes in the diversity of human forms alive today we know that we are one variable species. That is not the case in the fossil record, where around one-third of all dinosaur species known are represented by a single, almost always incomplete specimen. In some cases, like a spectacular specimen in the Hall of Ornithischian Dinosaurs at the Museum which preserves large skin patches, we cannot even determine what ceratopsian it is because it lacks a skull.

These Late Cretaceous ceratopsians from North America are so conservative in body form that without their ornate skulls they are nearly impossible to tell apart. This has significant implications for dinosaur biology. Classically, people who name species (taxonomists and systematists) have been pigeonholed into two camps: splitters and lumpers. Splitters name new species on the basis of the tiniest of variations. Lumpers over-account for variability and throw lots of very different specimens into the same species. From today's biological perspective it is still difficult to evaluate these things. For instance, we know that many birds, fish, and lizards look identical to one another from our visual perspective. Yet when we look at DNA sequences, they are clearly different species. This is a win for the splitters. Conversely, lumpers, citing the incredible variation similar to that seen in humans, tend to consolidate lots of variability into a single species. One can see how these differing philosophies impact the field of dinosaur science as that diversity can be way overestimated (splitters) or way underestimated (lumpers). Because it is so difficult to tell even living species apart and the fossil record is so poor, most contemporary paleontologists tend to split rather than lump. Hence it is highly likely that *Centrosaurus* and *Monoclonius* are in fact different species.

⌐ This specimen of *Monoclonius* even preserves the bones from inside the eye. Called scleral ossicles, such bones support the lens and are found in many extant animals.

∨ This image of the pelvis shows fossilized tendons, which would have supported the tail aloft when the animal was alive.

∧ Barnum Brown collecting a specimen in Alberta, Canada in the early 1900s. This picture (like several in this book), is a hand-coloured glass slide from the extensive photographic archives at the Museum.

STYRACOSAURUS ALBERTENSIS

LATE CRETACEOUS
DINOSAUR PARK FORMATION
WESTERN NORTH AMERICA

WHEN IT COMES TO CRANIAL ORNAMENTATION, CERATOPSIAN DINOSAURS RUN THE GAMUT: THERE ARE SPECIES WITH BUMPS, HORNS, CRESTS, FRILLS, PITTED SCULPTURING AND DEEP GROOVES. HOWEVER, ONE OF THE REAL OUTLIERS IS *STYRACOSAURUS*. ITS NAME, MEANING "SPIKED LIZARD", DERIVES FROM THE SERIES OF SPIKES AND HORNS THAT COVERED ITS HEAD. THESE INCLUDED A LARGE, RHINO-LIKE NOSE HORN, SPIKES THAT PROTRUDED FROM THE SIDE OF ITS HEAD AND, MOST CHARACTERISTICALLY, A FENESTRATED FRILL PERIMETER THAT WAS ORNAMENTED BY LARGE SPIKES.

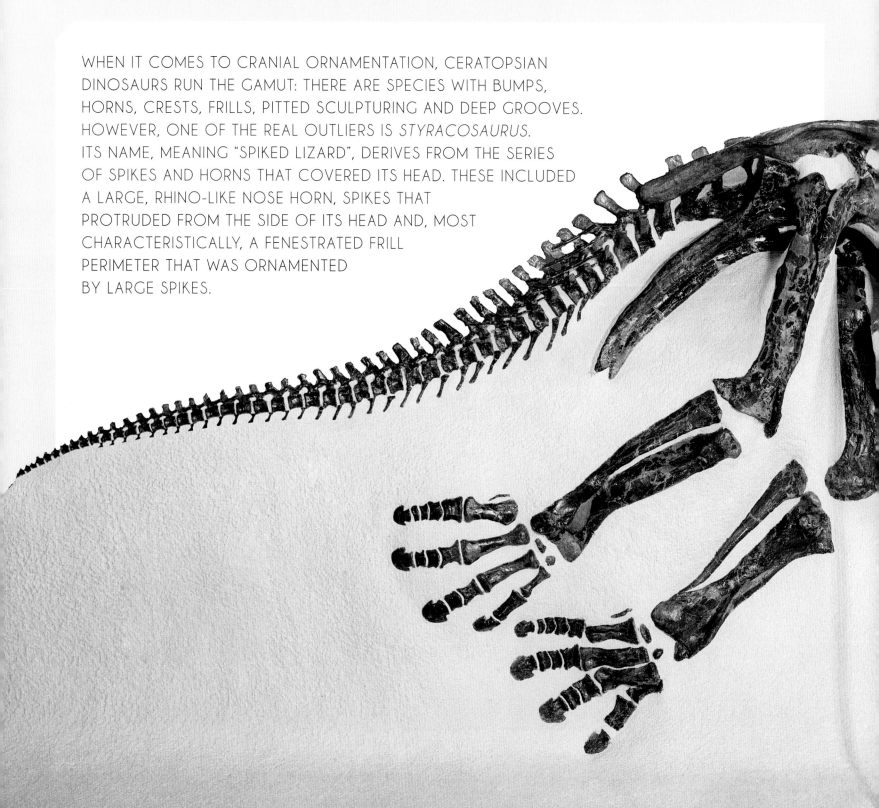

A fenestrated frill is one that is not solid; instead it is perforated by two large holes or windows, and differs from the solid frill in *Triceratops*. The spectacular crest spikes were as large as the nose horn.

Like other ceratopsians, *Styracosaurus* was a unwieldy quadruped. It was not exceptionally large, but even so at 5.5m (18ft) it would have weighed over 3 tonnes. It is thought to have

been a social animal because many bonebeds, some containing tens of individuals, have been found. Like almost all other ceratopsian dinosaurs, the use of the horns and frills is an object of interest, both in the professional and public sphere. Popular media and scientific journals alike have depicted these animals as posturing, displaying, fending off predators, sick with bone disease, dumping thermal

∨ *Styracosaurus* is an extraordinary-looking creature. This remarkable specimen shows it in all its glory. While it appears fearsome, the horns and spikes are now considered a display feature.

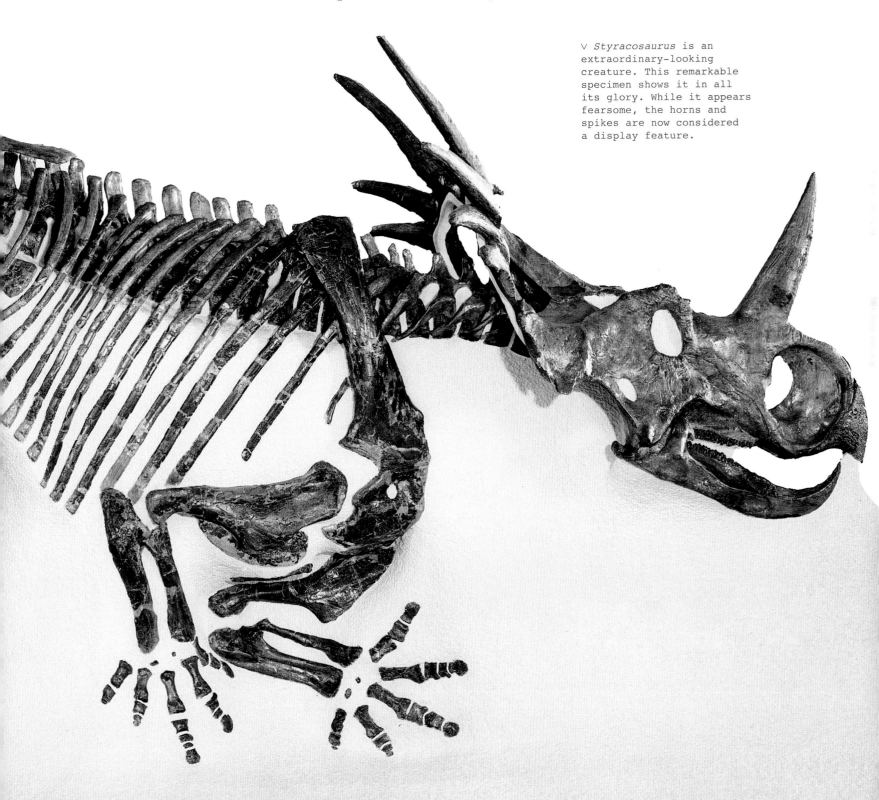

< In front view the big holes in the frill (called fenestrae) are apparent. This shows how fragile the skull would have been.

> *Styracosaurus* is one of the most heavily ornamented ceratopsian dinosaurs. Here it has a brightly coloured head. While we have no direct evidence for this, bright colours on extinct dinosaurs are expected.

↘ Museum technicians working on the *Styracosaurus* skull. Historically, skilled technicians have contributed as much to understanding and displaying dinosaurs as paleontologists themselves.

energy through their crests in the same way as elephants use their ears, or fighting among themselves. Even the purely pragmatic suggestion that the frills were used to support large jaw muscles has been proposed.

Detailed analyses seem to suggest that the horns and frills were not very sturdy. Assessment of the damage to the bones and frills indicates that it was not caused by thrusts of other ceratopsians. Healed wounds are rare, which also suggests that they were feeble defensive shields. Therefore, the consensus among professionals is that the frill was primarily a display structure.

Styracosaurus lived in a very diverse ecosystem. It competed with a host of herbivores, including duckbills such as *Corythosaurus* (see p.220) and *Lambeosaurus* among others, as well as closely related ceratopsians such as *Chasmosaurus* and *Centrosaurus*. Prevalent carnivores included *Gorgosaurus* and *Daspletosaurus*. At the time, this area of North America was much warmer than the present day with distinct wet and dry seasons. Because it was close to an intercontinental seaway that bisected North America, the climate was somewhat more maritime with warmer winters and cooler summers than continental interiors. The flora was composed primarily of conifers and ferns with a few angiosperms also present.

If we examine all of the herbivorous animals in this fauna we can detect subtle differences in feeding apparatus, such as shapes of jaws, the number and architecture of teeth, and so on. Just as today, when there are a number of similar animals in the same habitat, animals use these specializations to partition the available resources. This is known as niche partitioning, and it makes it possible for a large diversity of similar animals to occupy a single area. *Daspletosaurus* was much more heavily built than *Gorgosaurus*. This size difference was probably also reflected in the kinds of prey and feeding strategy that these two specialized animals utilized.

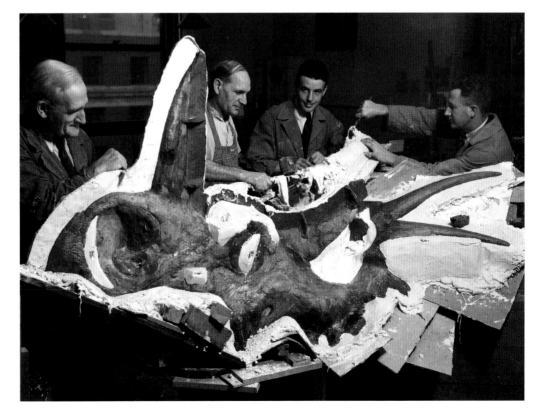

HYPSILOPHODON FOXII

EARLY CRETACEOUS
WEALDEN FORMATION
EUROPE

HYPSILOPHODON IS A SMALL (ABOUT 2M/6½FT) HERBIVOROUS DINOSAUR FOUND ONLY ON THE ISLE OF WIGHT, AN ISLAND OFF THE SOUTH COAST OF ENGLAND. ITS NAME IS IN REFERENCE TO ITS TEETH, WHICH RESEMBLED THOSE OF THE GREEN IGUANA (*IGUANA IGUANA*), PREVIOUSLY CALLED *HYPSILOPHUS*.

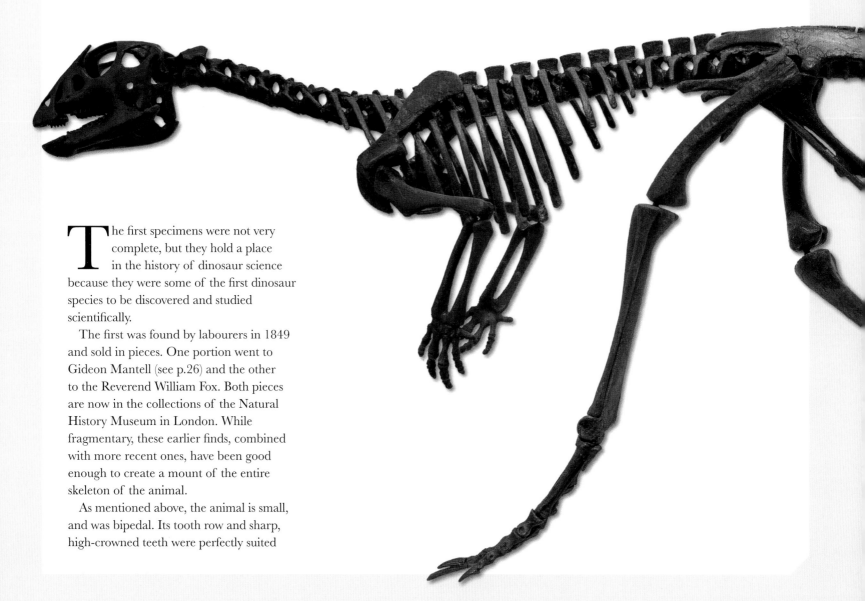

The first specimens were not very complete, but they hold a place in the history of dinosaur science because they were some of the first dinosaur species to be discovered and studied scientifically.

The first was found by labourers in 1849 and sold in pieces. One portion went to Gideon Mantell (see p.26) and the other to the Reverend William Fox. Both pieces are now in the collections of the Natural History Museum in London. While fragmentary, these earlier finds, combined with more recent ones, have been good enough to create a mount of the entire skeleton of the animal.

As mentioned above, the animal is small, and was bipedal. Its tooth row and sharp, high-crowned teeth were perfectly suited

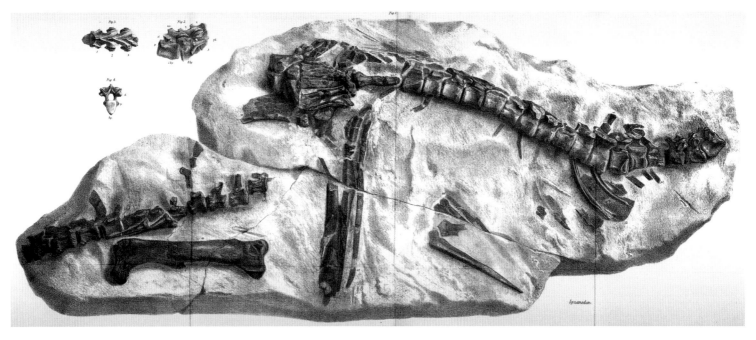

∧ *Hypsilophodon* fossils are not that rare in southern England. This is one of the first to be discovered and is referred to as the Mantell-Bowerbank block. It is the paratype for the taxon.

< *Hypsilophodon* is a primitive herbivore on the ornithopod line. It is represented by several specimens. However, they are only partially complete.

for chewing tough, fibrous plants. The end of the mouth terminated in a sharp beak, which was probably covered in keratin, like the beak of a modern bird. This would have facilitated plant harvesting before the food went on to be processed by the cheek teeth.

The first specimens of *Hypsilophodon* were discovered early in the history of dinosaur study, and in retrospect much of what had been presented has proved to be rather odd interpretations of its lifestyle. These include portraying it as quadrupedal; as an arboreal animal with feet that had opposable toes able to grasp branches; and even as an evolutionary mimic of a tree kangaroo. All of these have been debunked, and modern thought considers *Hypsilophodon* and its relatives to be relatively

typical dinosaurs with a bipedal gait, as is true of many small dinosaurs.

Although remains of *Hypsilophodon* are restricted to the Isle of Wight, close relatives in the "hypsilophodontid" group are more widely known. It is worth noting that "hypsilophodontid" dinosaurs are not a recognized group (see p.52), it is a "grade" of animals that lies near the base of the ornithopod dinosaurs. These "hypsilophodontids" have been found almost globally with the notable exception of Africa.

Some of these are exceptional for their geographic position. *Leaellynasaura* is one example. Its remains have been found at the Early Cretaceous Dinosaur Cove locality on the south coast of Australia. This is an

interesting area, as during the time that *Leaellynasaura* lived, it would have been well within the Antarctic Circle. *Leaellynasaura* was a rather small animal (1m/3¼ft), so it is unlikely it would have migrated long distances, and even though the climate was more temperate than today (there were no polar ice caps at this time), it still would have had to endure months of very low light conditions and cold temperatures. Some anatomical evidence has been proposed for this, as the optic lobes (the part of the brain that processes visual information) and the orbits (the area of the skull that houses the eye) seem to be large, indicating great visual

acuity. Others have challenged this idea, suggesting that these features are typical of many small dinosaurs.

Another "hypsilophodontid" is *Thescelosaurus*. It is a larger animal than *Hypsilophodon* or *Leaellynasaura* (large individuals are a little over 4m/13ft) in length), and lived in North America in the Late Cretaceous. Like all of these animals it was a bipedal herbivore, rather unremarkable in appearance. A very complete specimen, now in the collections of the North Carolina Museum of Natural Sciences in Raleigh, USA, was collected in the early 1990s. Given the familiar name "Willo", this specimen

had a large concretion in the middle of its chest. After preparation and CT scanning, the scientists managing the project made the incredible claim that this concretion represented Willo's fossilized heart; it was so well preserved that even fine details of anatomy could be examined. They went on to make a number of claims about how this anatomy related to presumed physiology and dinosaur relationships. At the time of publication, this announcement generated worldwide press attention. Several subsequent studies have shown this body not to be a fossilized heart, but simply a large mass that formed inside Willo's torso after death.

< The hypsilophodontid *Leaellynasura* was a small animal that lived in the Early Cretaceous of what is now southern Australia.

> The brown lump encircled by the ribs and bones of the shoulder in this *Thescelosaurus* specimen has been proposed as a fossilized heart. However, added research disputes this claim.

∨ An articulated specimen of *Thescelosaurus* on display at the Museum of Rockies in Bozeman, USA. *Thescelosaurus* was among the largest and last of the hypsilophodontids.

HYPACROSAURUS ALTISPINUS

LATE CRETACEOUS
HORSESHOE CANYON FORMATION
WESTERN NORTH AMERICA

UP UNTIL RELATIVELY RECENTLY THE DISCOVERY OF BABY DINOSAURS AND DINOSAUR EGGS WERE VERY RARE EVENTS. IN THE AMERICAN MUSEUM OF NATURAL HISTORY DINOSAUR HALLS, THERE ARE MANY BABY DINOSAURS ON DISPLAY, INCLUDING *PROTOCERATOPS*, *PSITTACOSAURUS* AND *HYPACROSAURUS*. *HYPRACROSAURUS* WAS A FAIRLY GENERALIZED HADROSAUR.

Like other lambeosaurs it had a large hollow crest on its head and was averaged sized as an adult (about 9m/29ft). Although it is not a common dinosaur, important specimens of eggs, embryos and juveniles have been collected. These tell us much about the physical changes in *Hypacrosaurus* as it grew, especially the development of the crest and muzzle. Examination of the embryos also indicate that it had a long incubation time, around

170 days in the egg before parturition. Specimens of this dinosaur have also been found in Late Cretaceous sediments in northern Montana.

Officially the study of baby dinosaurs took on its modern dimension in the 1970s, when a local rock shop owner showed some small bones to Jack Horner, who at the time was working at Princeton University. The bones had been found near the town of Choteau in Montana, and

Horner immediately recognized them as the remains of baby duck-billed dinosaurs, which were given the name of *Maiasaura* meaning "good mother lizard". After securing permission from the landowners, Horner began excavation at the site. Today this area is windswept prairie, with blistering hot summers and freezing winters. In the time of *Maiasaura*, however, it was much more tropical and populated by a rich fauna of dinosaurs, other

archosaurs such as crocodilians and birds, as well as lizards, turtles and mammals.

It is a very rich site, still under excavation, and new species and specimens are being found on almost an annual basis. The locality has been called Egg Mountain because so many dinosaur eggs and nests have been found there, and is fairly typical of other hadrosaur nesting grounds, except it is much better preserved.

A great deal is known about *Maiasaura* from the extensive sample at Egg Mountain. From analysis of the plentiful juvenile dinosaur remains, it has been concluded that a very large percentage (perhaps nearly 90 per cent) perished in their first year. They also tripled their size to almost 0.5m (1½ft) during that time, and purportedly growing to adult size in eight years. When they reached adulthood the picture did not look a lot brighter, with mortality spiking at about 50 per cent just a few years after the animals were fully grown. At Egg Mountain a large number of nests have been found, and many scientists have concluded that the animals

nested communally. The nests themselves are hollows about 1m (3¼ft) across, probably excavated by the mother, and contain 30–40 eggs, each one about the size of an ostrich egg; *Hypacrosaurus* eggs are about the same size and shape. Unlike advanced theropod dinosaurs, such as *Citipati* (see p.98), it is not thought that they brooded their eggs. Instead they probably covered them with vegetation, which provided heat for the developing eggs as it rotted. This is the strategy used by alligators today.

Some have suggested that because such a high proportion of the bones are not fully coossified (in both *Maiasaura* and *Hypacrosaurus*) many of the babies were so fragile that they could not walk for a substantial time after hatching. This may mean that adults would have had to bring food back to the brood for them to survive as they were incapable of foraging for themselves. This is not universally accepted, as many hatchling animals today do not have fully ossified bones but are still perfectly capable of walking.

∧ Much has been suggested about parental care in *Hypacrosaurus* and other hadrosaurs. To determine this definitively much more work needs to be done.

< One of the dinosaurs for which we have good ontogenetic (growth) information is *Hypacrosaurus*. The age range from embryos to adults are known from fossils.

EDMONTOSAURUS ANNECTENS

LATE CRETACEOUS
HELL CREEK FORMATION
NORTH AMERICA

EDMONTOSAURUS WAS ONE OF THE MOST COMMON DINOSAURS TO ROAM WHAT IS NOW THE WESTERN PLAINS OF NORTH AMERICA AT THE END OF THE CRETACEOUS PERIOD. LARGE HERBIVORES BELONGING TO THE HADROSAUR GROUP, THESE DINOSAURS ARE OFTEN CALLED DUCKBILLS, BECAUSE THE END OF THEIR SNOUT IS SHAPED LIKE THE END OF A DUCK'S ROSTRUM. THIS BEAK WAS PROBABLY AT LEAST PARTIALLY COVERED WITH HORNY KERATIN IN A SIMILAR WAY TO THOSE OF TURTLES AND BIRDS TODAY.

Although hundreds of *Edmontosaurus* specimens are known, only a small number are adults. This is true of most dinosaurs: the majority were still growing when they died, which indicates that pre-adult mortality was very high. An *Edmontosaurus* specimen on display at the American Museum of Natural History is a very large hadrosaur mounted in an upright pose, but in fact this posture is wrong. The specimen was one of the earliest hadrosaurs ever to be mounted. However, the upright posture requires the tail vertebrae to be disarticulated to such an extent that this pose would have been anatomically impossible in life. We now know that hadrosaurs (or at least most of them) were primarily quadrupedal in life and held their tails parallel to the ground and at the same level as their bodies. The head rose from the body on a sinuous-shaped neck.

Most of the modifications to the skull of these animals is directly involved with feeding. *Edmontosaurus* and some of its relatives have a huge number of teeth – hundreds of them. These are organized into dental batteries that form broad, flat chewing surfaces, which are perfect for orally processing huge amounts of high-quality food for digestion. This is the complete opposite to the process in sauropods, where vegetal matter wass bulk processed and presumably fermented in the gut.

There are dinosaur fossils – and there are dinosaur fossils. One of the most spectacular fossils in the world is in the collection and on display at the Museum, and it is a specimen of *Edmontosaurus*. This particular specimen was collected by the Sternbergs, a family of American–Canadian fossil hunters who turned paleontology into a business. The work of the family's patriarch, Charles H. Sternberg, was alongside legendary paleontologist E.D. Cope in the late 1870s. He began academic study at Kansas State University, but never matriculated, as

> *Edmontosaurus* is a Late Cretaceous dinosaur from western North America. During its time it was probably the most common large herbivore in its environment.

his true passion was collecting fossils rather than describing them. The Sternberg legacy continued with his sons' participation, one of whom went on to be an important organizer of what would become Dinosaur Provincial Park, now a UNESCO World Heritage Site, in Alberta, Canada. In competition with academic paleontologists from museums and universities, the Sternbergs collected many important specimens. Several are still on display in museums across the world.

The *Edmontosaurus* specimen is often referred to as the "dinosaur mummy". It was collected near Lusk, Wyoming, USA, in the eastern part of the state. When Sternberg found it, he knew exactly how exceptional it was, even remarking that he was so excited that he could not sleep the night of the discovery. It was not a real mummy, in the sense that it was embalmed and dried. Instead, the specimen was desiccated – it had dried out in an extremely arid environment.

Such things still happen, and cattle, camel, zebra and elephants resembling deflated bags of bones covered with tough leather can be observed in desert and semi-desert today. Even a dead mouse may appear like this when found in a cupboard or under a stove. This is exactly what happened to the *Edmontosaurus*, except that after it dried out it was buried in soft sediment. The "skin" that is preserved is an impression rather than the skin itself. It shows a pattern of tubercles or bumps that are of two types, large and small. The large tubercles are arranged in clusters and surrounded by a matrix of smaller tubercles. The skin also shows wrinkles, like the flappy, loose skin around the legs of a rhinoceros or elephant. This indicates that the skin was elastic and flexible and would accommodate a broad range of movement. The forefeet show that the hands sported hoof-like structures, which correspond with the trackways known for these animals. The hooves were probably keratinous in structure like horse's hooves today. This specimen is one of the crown jewels of the Museum's collection, and still delights at every viewing.

⌐ Charles R. Knight's reconstruction of a hadrosaur. This early reconstruction makes the mouth look more duck like than we think today.

∟ Charles H. Sternberg and his sons excavating the "mummified" *Edmontosaurus*. This would become one of the most treasured0 specimens in the Museum's dinosaur collection.

∧ Hadrosaur tooth rows are composed of dozens of teeth cemented together into a dental battery. This dental battery forms chewing surfaces that are replenished with new teeth arising from below with wear.

∨ The *Edmontosaurus* mummy. The skin is so finely preserved that the individual tubercles and the hoof-like pads on the forefeet can be easily observed.

TENONTOSAURUS TILLETTI

LATE—EARLY CRETACEOUS
CLOVERLY FORMATION AND OTHERS
WESTERN NORTH AMERICA

TENONTOSAURUS IS A PRIMITIVE BUT INTERESTING ORNITHOPOD DINOSAUR FROM WESTERN NORTH AMERICA. IT IS CONSIDERED TO BE ONE OF THE MAIN HERBIVORES OF ITS ENVIRONMENT, AS IT IS THE MOST COMMON VERTEBRATE FOSSIL FOUND IN THE SEDIMENTS WHERE IT IS ENCOUNTERED.

Known from the middle part of the Cretaceous, about 110 million years ago, most of the specimens come from Montana and Wyoming in the USA, with a few collected in adjacent areas, including one from Maryland on the Eastern Seaboard. *Tenontosaurus* was a medium-sized dinosaur, up to 8m (26¼ft) in length, and very conservative in morphology.

Like so many of the dinosaurs covered in this book, the first specimens of *Tenontosaurus* were collected by Barnum Brown. He found the first specimen in 1903, and several others were excavated by him in the 1930s. He informally named the animal "*Tenatosaurus*", which roughly translates as "sinew lizard". He never formally published his work as

the pressures of his career, global economic depression and the Second World War took their toll. The work was later picked up by John Ostrom of Yale University, who had studied at the Museum of Natural History as a graduate student. In 1970, he gave the specimen its official name *Tenontosaurus*. Inexplicably he never mentioned Brown's contributions to the study.

Brown's original reference for the name "sinew lizard" refers to the intricate system of imbricate tendons, which interlace like basket-weaving at the top of the tail and trunk vertebrae. This is powerful evidence that these animals had very stiff spinal columns and that, instead of dragging their tails along the ground like a crocodile, they

held them out behind the body, parallel to the ground. The tail may have had a reasonable amount of lateral flexibility, but very little dorsoventrally. The rigidity created by this truss system of ligaments is similar to a suspension bridge or cantilevered architecture. The tail, as in other dinosaurs, provided a heavy counterbalance to the torso; it may have assisted in bipedal locomotion, but certainly kept the animal from falling forwards while it was ambulatory.

There are some very well-preserved specimens of *Tenontosaurus*. Some of these appear to show that the animals were predated on by the dromaeosaur *Deinonychus* (see p.108). This evidence comes from the association of *Deinonychus* remains with those of *Tenontosaurus* at several independent sites. Several paleontologists have suggested that it is unlikely that an adult *Deinonychus* could have predated independently on an adult *Tenontosaurus* and go on to propose that this is

evidence that *Deinonychus* hunted in packs. In today's world adult lions, if they are famished, will occasionally hunt African elephants, but almost always in packs. It has also been suggested that *Deinonychus* preferred to hunt juvenile *Tenontosaurus* as opposed to large adults, and that it is possible *Deinonychus* may have cannibalized one another in a feeding frenzy during these events. All of this is based on the common occurrence of bones at the same sites, so it is speculative at best.

The flora and fauna associated with *Tenotosaurus* is interesting because it is transitional. Early in its history the climate was considered to be warm and arid, gradually shifting over a few millions of years, into a warm and humid environment. Associated fauna is depauperate, either because diversity was not high, or that the fossil record is extremely poor in these formations. There is scant evidence, for instance, of a predator larger than *Deinonychus*, which is unusual, but

some fairly large sauropods are present. There is still a lot to learn about *Tenontosaurus* in term of both its environment and paleoecology.

< While often not preserved and extremely delicate, the tails and dorsal vertebrae of most ornithopod dinosaurs were supported by interwoven tendons. These allowed the tail to be held aloft, cantilevered and parallel with the ground.

∟ In some instances, *Tenontosaurus* bones have been found associated with the isolated teeth of the carnivorous *Deinonychus*. This has led to speculation that *Deinonychus* hunted this animal in packs.

∨ A skeleton of *Tenontosaurus* on display in the Museum dinosaur halls. It is mounted in an old pose. A modern one would have the tail sticking straight back from the body.

SAUROLOPHUS OSBORNI

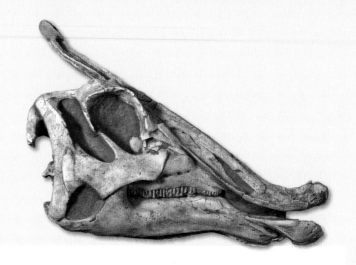

LATE CRETACEOUS
HORSESHOE CANYON FORMATION
WESTERN NORTH AMERICA

UNLIKE *CORYTHOSAURUS* (SEE P. 220) *SAUROLOPHUS* IS A SAUROLOPHINE HADROSAUR, AND LIKE *EDMONTOSAURUS* HAD A SOLID CREST ON ITS HEAD AS OPPOSED TO THE HOLLOW CREST OF LAMBEOSAURINES, SUCH AS *CORYTHOSAURUS* AND *LAMBEOSAURUS*. *SAUROLOPHUS* WAS A LARGE DINOSAUR AND A CONTEMPORARY OF *HYPACROSAURUS* AND *EDMONTOSAURUS*.

Saurolophus osborni was originally discovered in the Horseshoe Canyon Formation of Alberta, Canada. It was fairly large, about 10m (33ft) in length. Like other hadrosaurs, it was primarily quadrupedal as an adult, and probably partially bipedal as a juvenile.

A second species, *Saurolophus angustirostris*, has been found in Mongolia. It is somewhat larger than the North American variety, with a skull up to 20 per cent longer than *Saurolophus osborni*. Multiple specimens of this hadrosaur were found by Russian paleontologists at the colourfully named Tomb of the Dragons locality in the Nemegt Basin of Mongolia. After paleontologists from the American Museum of Natural History ceased working in Mongolia due to political instability and conflict in Central Asia during the early 1930s, the rich fossil localities of the Gobi Desert lay fallow until Soviet paleontologists began working in the area. They were far better equipped than the earlier Museum expeditions, which allowed them to venture much further out in the desert than the Americans. Among their many discoveries were the rich Late Cretaceous dinosaur-bearing rocks of the Nemegt Basin. In addition to *Saurolophus*, these rocks produced spectacular specimens of many Late Cretaceous dinosaurs, including

Tarbosaurus (a close relative of *Tyrannosaurus rex*), ostrich dinosaurs like *Gallimimus*, several ankylosaurs, and pachycephalosaurs. These beds are still being actively excavated.

As in most hadrosaurs, the head of *Saurolophus* was crowned with an ornament.

The same arguments regarding function have been applied here as for other hadrosaurs. Some have suggested the crest was for the production of sound; it is not hollow itself, as in lambeosaurines, but there was a shallow hollow at its base.

While there is no definitive way to determine the function of these crests, the popular consensus is that they were used for display or species recognition.

Saurolophus is one of the few dinosaur taxa that have been found on two continents. This is not that surprising, because at the time this animal lived the fauna and climate of this part of Asia and Western North America were very similar, and there existed an intermittent land bridge between North America and Siberia which allowed faunal interchange. Because *Saurolophus* is most closely related to North American hadrosaurs, it is thought that it migrated across this land bridge – Alaska and Siberia lay along the same latitudes during the Late Cretaceous as they are today, but the climate was much warmer and could support a rich dinosaurian and mammal fauna. How these animals dealt with near darkness during the winter is hard to determine. Some have suggested that they migrated to more southerly areas to avoid the winter, but since very small animals are present it is unlikely that these were capable of seasonal migration over long distances.

⌐ *Saurolophus* is an unusual dinosaur in that it is found both in the Late Cretaceous of North America and in Central Asia.

∨ The Museum's *Saurolophus* specimen is one of the jewels of the collection and is almost perfectly preserved.

∧ Barnum Brown excavating the Museum's *Saurolophus* specimen in Alberta, Canada.

CORYTHOSAURUS CASUARIUS

LATE CRETACEOUS
DINOSAUR PARK FORMATION
ALBERTA, CANADA

AS THE SPECIES NAME IMPLIES, THE OBSERVATION THAT
THIS DINOSAUR HAD AN ELABORATE HOLLOW HEAD CREST
THAT IS VERY SIMILAR TO THE EXTANT CASSOWARY WAS
IMMEDIATELY OBVIOUS. THERE ARE TWO GROUPS OF ADVANCED
HADROSAURS, THE HADROSAURINES AND THE LAMBEOSAURINES.

The latter is characterized by their elaborate head crests. As with so much else in these chapters, much has been made of the function of this elaborate anatomy, but very little is actually testable within a scientific framework.

What the crest of *Corythosaurus* was actually used for is very difficult to determine. Even in the cassowary, the functionality of the crest and the evolutionary pressures it is under are open to question. It has been suggested that it was a display feature, that it was used to enable species recognition, to produce low-frequency sounds, or that it was used

for ploughing through dense vegetation.

The casque head of *Corythosaurus* was impressive. It comprised a large, hollow crest of bone reminiscent of a Corinthian battle helmet from classical antiquity, which gave the genus its name. One factor that suggests the crest was for display is that it did not start developing until the animal was about half grown; this is typical of many secondary sexual characteristics in mammals today, such as horns. Several other suggestions have been proposed, although all are difficult to test. One compelling and popular theory is that the hollow crests worked as sound

amplifiers for communication via low-frequency sound. Supporting evidence may be provided by the complexity of the ear in these dinosaurs. The excellent sample we have of very complete skulls

> Many reptiles with large crests on their heads like *Corythosaurus* exhibit brilliant colours. This reconstruction extends this concept to these dinosaurs.

∨ This specimen of *Corythosaurus* is so remarkably preserved it has numerous impressions of the skin on the body.

∧ The *Corythosaurus* crest
was hollow, leading some to
speculate that it may have
been a resonating chamber to
produce sound.

> The pebbly skin texture
of *Corythosaurus*. In life,
as portended by the visible
folds, the skin was soft and
pliable.

facilitates this in comparison with most other dinosaurs, and at the very least studies have shown that the ears of *Corythosaurus* may have been highly sensitive.

Corythosaurus specimens were first excavated in 1911 by Barnum Brown. Found in Late Cretaceous sediments along the Red Deer River of what is now Dinosaur Park in Alberta, Canada, this animal was a little over 8m (26¼ft) in length and was probably primarily quadrupedal. Stomach contents consisting of conifer parts have been recovered. The type specimen (on display at the Museum) preserves several pieces of skin, showing (at least these areas) that it was covered with small polygonal scales. This specimen also shows a characteristic that is extremely common in dinosaurs, both extinct and extant. Almost all dinosaurs have fairly

long, flexible necks (at least in relation to the necks of mammals, crocodiles, lizards and turtles). When a long-necked animal dies, the muscles that hold the head down with the eyes focused forwards or to the ground relax, and the head is then pulled back by strong, desiccated tendons. This is called a death posture and is the position in which many famous dinosaur specimens have been found. You can observe the same posture in dead birds that you might occasionally encounter along the shoreline in coastal areas: most of the time the head will be snapped back above the neck as in many non-avian dinosaur fossils.

The depredations of human conflict have also dealt a mighty blow to our knowledge of *Corythosaurus*. The two best specimens ever collected were excavated by noted private fossil collector Charles H. Sternberg in

the same beds from which Barnum Brown collected the Museum specimens. Sternberg found these in 1912, at a time when Brown and Sternberg were involved in what was usually (but not always) an amicable competition for dinosaur bones in this area of Canada. Sternberg had arranged for the British Museum of Natural History (as it was called at the time) to purchase the finds. The specimens were packed up and sent to Keeper of Geology Arthur Smith Woodward on board the SS *Mount Temple* bound for London in 1916. (The *Mount Temple* is noteworthy as one of the ships that assisted in the rescue of passengers from the *Titanic* in 1912.) Sadly, the *Mount Temple* was sunk on its way to London by a German U-boat, resulting in the loss of four merchant seaman and the precious dinosaur cargo.

BIRDS AND DINOSAURS

A COMMON THREAD RUNNING THROUGH THIS BOOK IS THAT LIVING BIRDS ARE A TYPE OF DINOSAUR. IF THIS BOOK HAD BEEN WRITTEN 25 YEARS AGO THIS WOULD HAVE BEEN A CONTENTIOUS STATEMENT, EVEN THOUGH THE IDEA HAD FIRST BEEN PROPOSED OVER 100 YEARS EARLIER. IN THE INTERIM ALL KINDS OF NEW EVIDENCE HAS BEEN ASSEMBLED IN SUPPORT OF THIS IDEA – TO THE EXTENT THAT THE THEORY IS NOW NEAR-UNIVERSALLY EMBRACED.

Character support for this relationship began with Thomas Huxley's observation that *Archaeopteryx* shared many characteristics with non-avian dinosaurs. Huxley's ideas were rekindled in the 1960s, largely by Yale paleontologist John Ostrom who had intensively studied *Archaeopteryx,* as had previous generations of paleontologists. But in addition to this, he had excavated more specimens of the dromaeosaur *Deinonychus,* including a fair amount of skull material. His analysis of these finds pointed to even more characteristics that advanced theropod dinosaurs (such as *Deinonychus*) had in common with primitive birds. He even advanced the idea that if *Archaeopteryx* had been found without feathers it would have been considered a non-avian dinosaur rather than a bird. To many biologists and paleontologists, birds seem almost to have been specially created for powered flight, and most of their defining characteristics were considered as adaptations for this purpose. Hollow bones and a lack of teeth reduced weight, feathers provided an airfoil, and the advanced lungs fostered the high metabolism required to sustain flight.

In the world of systematic biology and paleontology the 1970s was an exciting time. Scientists, including many at the American Museum of Natural History, developed a new method called cladistic analysis. Previously, the study of genealogy was a largely subjective affair, with family trees drawn simply based on the experience and preferences of the

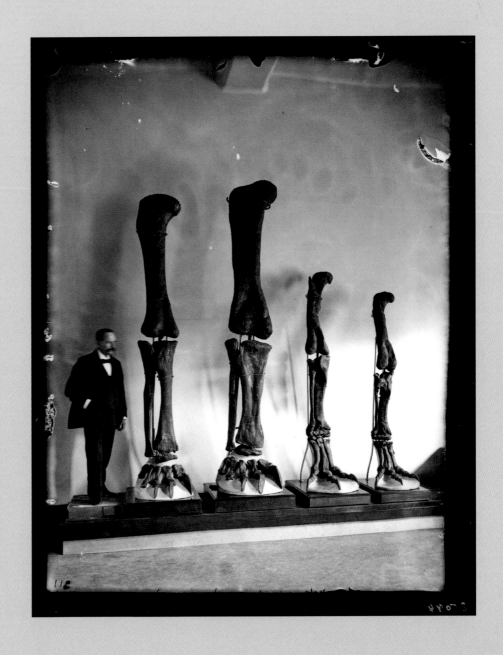

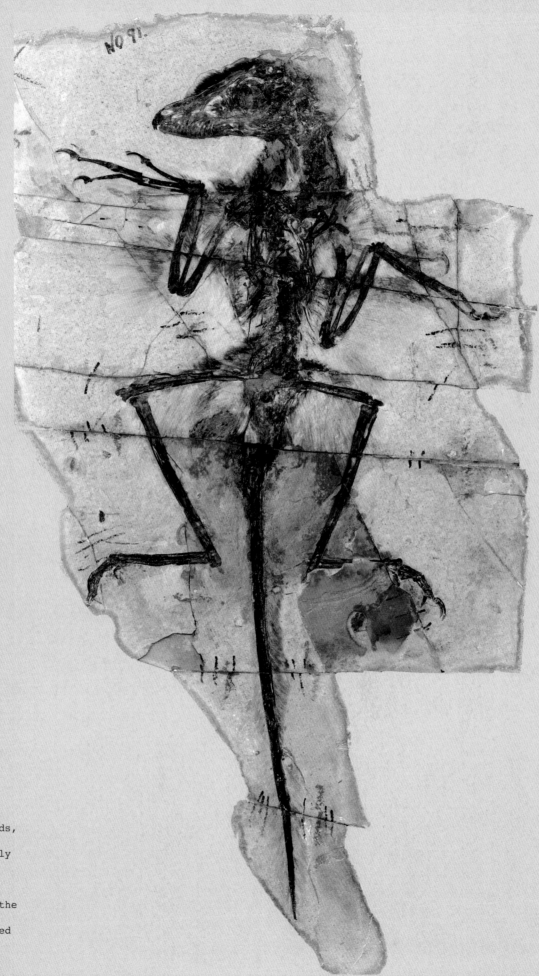

< Of these four dinosaur legs, the two on the right (*Allosaurus*) are obviously more closely related to birds, as their feet with three forward pointing toes clearly indicates.

> The *Sinornithosaurus* specimen "Dave" was one of the first feathered dinosaurs to be announced, and it received global attention.

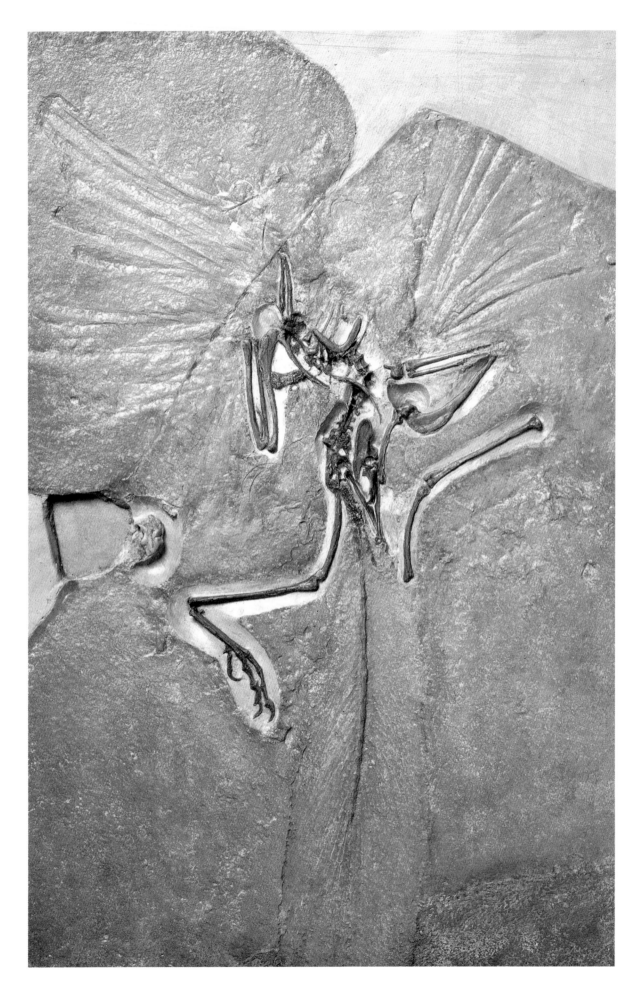

investigator. Cladistic analysis made the estimation of phylogeny an empirical pursuit, and provided a method where sets of characters were concatenated into a matrix. This matrix was then analyzed using specific computer programs which calculated the smallest number of evolutionary steps required to explain the changes in characters seen in the matrix.

The first to apply this to the question of bird origins was Jacques Gauthier while he was a graduate student at the University of California, Berkeley. He analyzed characters and found that *Archaeopteryx* and other birds shared a common ancestor more recent than any non-avian dinosaurs. This

relationship was conventional wisdom and had been known almost since the day the first *Archaeopteryx* was discovered. He called this group Avialae. His result also demonstrated that the closest group to Avialae was a group he called Deinonychosauria (see p.53), which included dromaeosaurs (such as *Deinonychus* and *Velociraptor*) and troodontids (*Saurornithoides* and *Mei long*). Ostrom was right: not only was *Deinonychus* similar to birds, it was one of the closest non-avian relatives of this group.

The conclusive evidence began to appear in the 1990s, when amazing animals were found in Mongolia and China. Many of the Chinese specimens displayed feathers and both

∧ A *Velociraptor* specimen collected in the Gobi Desert. The wishbone is clearly preserved as the V-shaped element in the middle of the figure.

< The London Specimen of *Archaeopteryx*. This was the specimen that spiked Huxley's interest in the relationship between theropod dinosaurs and birds.

avialans and feathered non-avian dinosaurs were found in the same deposits. In Mongolia, animals with extreme birdlike behaviours were found, including animals sitting on their nests brooding their eggs.

Over the years new discoveries and new types of analyses have appeared, which have pushed nearly every characteristic that was thought to have evolved in birds back to an origin far earlier in the dinosaur tree. Characteristics such as hollow bones, wishbones, an advanced metabolism, feathers and large brains were all present in dinosaurs long before birds and powered flight evolved. What this means is that dinosaurs developed all of these adaptations for some reason other than flight. It also has dramatic implications for the appearance of non-avian dinosaurs. If we could transport ourselves back to an Early Cretaceous forest in what is now Northeastern China, I think most of us would say, "Wow, look at all those strange-looking birds!"

∧ *Deinonychus* mounted in an active lunging pose.

> The skeleton of a modern galliform bird. Except for the lack of teeth and a long tail, the bones are extraordinarily similar to the *Deinonychus* on the facing page.

EXTINCTION

WHY DID DINOSAURS BECOME EXTINCT? THAT IS THE MOST COMMON QUESTION PALEONTOLOGISTS GET ASKED. IT WAS ALSO ONE OF THE FIRST QUESTIONS POSED WHEN SCIENTISTS STARTED TO STUDY THESE ANIMALS IN A FORMAL WAY IN THE NINETEENTH CENTURY. THIS IS A VERY COMPLEX PROBLEM, WITH MYRIAD VARIABLES. AS WOULD BE EXPECTED OF SOMETHING SO COMPLEX, MANY EXPLANATIONS HAVE BEEN PUT FORWARD.

These run from the comprehensible, such as increased competition from mammals, catastrophe, or slow climatic change, to the silly and even absurd – extreme constipation caused by the evolution of new plants, overhunting by space aliens, lack of sex drive, or a caterpillar plague that devoured all the plants, to list a few. Little time will be spent on the second category, but because this is such a difficult question, the former requires some analysis.

The easy response to why the dinosaurs became extinct is that they did not. As I have reiterated throughout the book, and explained in more depth in the previous chapter, living birds are a kind of dinosaur. I know what you are thinking – that pigeons can't really be compared to animals like *Tyrannosaurus rex*, *Triceratops*, and *Thescelosaurus*, all of which

> What the world may have looked like after the Cretaceous–Paleogene apocalypse.

∨ A rendering of the asteroid impact off the Yucatan peninsula at the end of the Cretaceous period, about 65 million years ago.

were exterminated at the Late Cretaceous extinction event. But that is not the point. Animals like *Microraptor*, *Velociraptor*, *Saurornithoides* and *Oviraptor* were in most ways more similar to modern birds than they were to other non-avian dinosaurs. They had feathers of modern aspect, they brooded their nests, their eggs had the same ultrastructure, and they had similar brains.

In fact, you could easily make a case that we live in the age of dinosaurs. What is meant by this? On Earth today among backboned terrestrial animals there are a little over 5,000 species of mammals, about 8,000 non-avian reptiles, and around 6,000 amphibians. It has been proposed that there are 18,000 species of birds – more than the number of mammals and non-avian reptiles put together. Not only are dinosaurs alive, they are doing very well.

When people generally think of dinosaurs they don't mean birds – even though they should. They mean what are improperly called "real dinosaurs".

It is obvious that something happened at the end of the Cretaceous Period about 65 million years ago that disrupted life on this planet. This great extinction was noted early in the history of paleontology and affected all habitats. It is most apparent and best studied in marine rocks. At this extinction event many important groups disappear. Marine animals such as ammonites, reef-building clams, sea-going reptiles like plesiosaurs and mosasaurs, and many microscopic animals became extinct at this time. Some estimates suggest that as many as 75 per cent of Earth's species may have disappeared. This is only second in magnitude to the Permian–Triassic extinction event which

∧ This is a sediment sample from the Hell Creek Formation. The grey layer signified by the arrow is the ash layer laid down by the impact.

∨ An image from space of the Yucatan peninsula in Mexico. The faint green curve on the upper left is a geological remnant of the impact crater.

> Non-avian dinosaurs were not the only animals to become extinct after the impact. Ammonites (relatives of the chambered nautilus) also completely disappeared at this time.

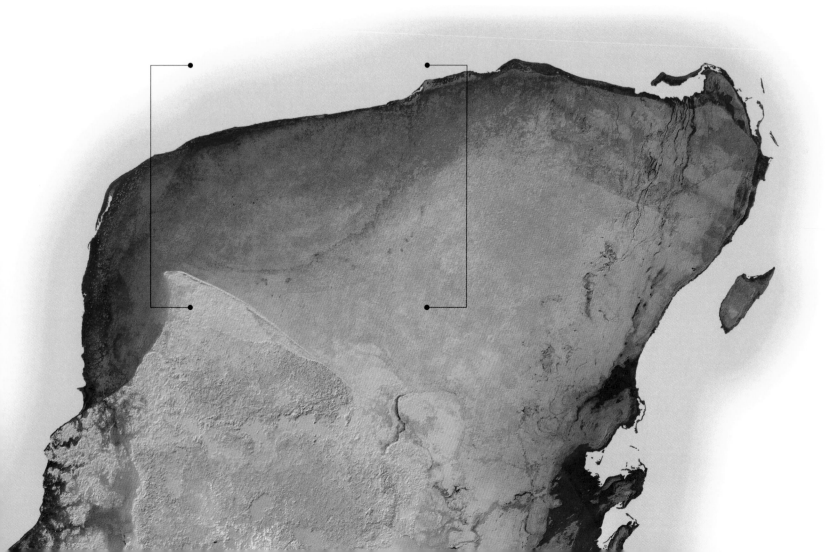

is thought to have seen over 90 per cent of animal and plant species wiped out.

This was a dramatic event in Earth history. Exactly what happened? Over the last few decades it has become apparent that a large extraterrestrial body collided with Earth. The site of the impact is just off Mexico's Yucatan peninsula, where the remnants of a large crater are still visible in seismic profiles of the ocean floor. The size of the object was 6–9 miles (10–15km) in diameter. This means that when it hit the planet its outer edge was near the upper limit of the troposphere. When it slammed into Earth, it vaporized and caused a dust plume of ejecta and debris that spread around the planet distributing a layer of dust. This, in fact, is how the event was discovered.

In Italy there are historic clay deposits which have been used for millennia to craft bricks and pottery. In the late 1970s a well-known physicist was contributing to a project to geochemically fingerprint the clay to determine if it was the source of raw materials for artifacts found throughout the Italian peninsula and adjoining areas. In one of the layers the team found that there was an exceedingly strong concentration of the element Iridium. Iridium is a very rare element on the Earth's surface. It is a common part of planetary interiors, comets and asteroids. Since this layer was in the exact stratigraphic position between strata of the Mesozoic era and later periods (the point of the mass extinction), could it have had something to do with the disappearance of so many species worldwide? Further analyses at the boundary in other locations showed further evidence, including an ash layer with high Iridium concentration, shock-fractured quartz (little pieces of the mineral which had been deformed by tremendous pressure) and microtektites.

Microtektites are small glass spherules which were components of the ejecta from the impact that melted as they re-entered the atmosphere, and are distributed around the globe. Certainly this all pointed to what is now considered fact – that a giant extraterrestrial body collided with the planet about 65 million years ago coincident with the mass extinction.

The immediate effects of the impact are controversial. The amount of energy that would have been released is almost unfathomable – over a billion times greater than the Hiroshima atomic bomb. Unarguably it was a global catastrophe. All sorts of phenomena have been proposed as direct effects of the impact. These include global firestorms, nuclear winter, acid rain, intense infrared radiation, and other chemical changes to the atmosphere.

The fossil record shows that life was severely affected. In the oceans, much of the plankton

∧ The extensive badlands of the Hell Creek Formation in eastern Montana and western South Dakota is the most heavily studied terrestrial impact system.

< About the same time as the impact, immense volcanic activity was also occurring, especially in what is now India. This released millions of tons of greenhouse gasses and ash into the air and formed the Deccan Plateau.

died and the record in some places indicates that the waters went from rich productive communities to diversity deserts. On land many of the plants were destroyed and the first to bounce back were ferns. This is aptly called the "fern spike", and it is a global phenomenon where the spores of ferns are the most common terrestrial plant remains. The largest terrestrial animals were under 25kg (56lb), and many groups disappeared completely. All of the larger dinosaurs and all dinosaurs except for the birds became extinct.

But does the collision explain the entire extinction? Probably not. Many other factors were in play. The Indian subcontinent (although then not part of Asia) was the centre of some of the largest volcanic eruptions in Earth's history. These occurred for over a million years surrounding the boundary and would have had profound climatic effects. Furthermore, climates were also changing

because the large intercontinental seaways that bisected the middle of North America and Asia were drying up. This has the effect of making continental interiors more seasonable. Today cities on or near coastlines have milder climates than cities in the middle of a landmass. There has even been the suggestion that there were several impacts rather than a single one. All of these could have contributed to the dinosaurs' demise before the final "coup de grace" of the Yucatan collision.

The major roadblock to gaining a better understanding of this is a lack of good samples, especially in the terrestrial record. There are only a handful of localities in the world where there are both dinosaur fossils and the impact layer, hence we do not have a complete picture of the events culminating in the impact or what happened immediately afterwards. Hopefully such sequences will be found somewhere in the world.

CREDITS

American Museum of Natural History
6 Photo Studio/D. Finnin/© AMNH, 9 Photo Studio/D. Finnin/© AMNH, 15 Photo Studio/D. Finnin/© AMNH, 16 Research Library/Image 17808, 17 Photo Studio/R. Mickens/© AMNH, 19 Research Library/Image 284863, 22-23 Photo Studio/D. Finnin/© AMNH, 25 Research Library/Image 275349, 30 Photo Studio/D. Finnin/© AMNH, 31 Department of Vertebrate Paleontology/Image CM1068, 34-35 © Mick Ellison, 36-37 Photo Studio/D. Finnin/© AMNH, 39 Photo Studio/D. Finnin/© AMNH, 40 Photo Studio/D. Finnin/© AMNH, 43 Photo Studio/D. Finnin/© AMNH, 44 Department of Vertebrate Paleontology/M. Ellison/© AMNH, 45 Department of Vertebrate Paleontology/D. Barta/© AMNH, 47 Photo Studio/D. Finnin/© AMNH, 49 Photo Studio/D. Finnin/© AMNH, 51 Department of Vertebrate Paleontology/M. Ellison/© AMNH, 52 (bottom) Department of Vertebrate Paleontology/M. Ellison/© AMNH, 56 Department of Vertebrate Paleontology Archives, 58 Department of Vertebrate Paleontology/M. Ellison/© AMNH, 59 (top) Department of Vertebrate Paleontology/M. Ellison/© AMNH, (bottom) Department of Vertebrate Paleontology/A. Turner/© AMNH, 66 (top) Photo Studio/D. Finnin/© AMNH, (bottom) Research Library/Image 202, 67 (top) Research Library/Image 35047, 69 Research Library/Image 128003, 70-71 Research Library/Image 7758, 71 (bottom) Research Library/Image 5418, 72 Research Library/Image 18171, 74 Research Library/Image 18172, 75 Department of Vertebrate Paleontology, 76-77 Photo Studio/D. Finnin/© AMNH, 78 (top) Research Library/Image 310100, (bottom) Photo Studio/D. Finnin/© AMNH, 84 (bottom) Department of Vertebrate Paleontology/M. Ellison/© AMNH, 88-89 Photo Studio/D. Finnin/© AMNH, 93 (right) Research Library/Image ls6_8, 95 (top) Department of Vertebrate Paleontology Archives, 96 (top) Department of Vertebrate Paleontology/M. Ellison/© AMNH, (bottom) Department of Vertebrate Paleontology/M. Ellison/© AMNH, 97 Department of Vertebrate Paleontology/M. Ellison/© AMNH, 101 Department of Vertebrate Paleontology/M. Ellison/© AMNH, 102 (bottom) Department of Vertebrate Paleontology/M. Ellison/© AMNH, 104 Department of Vertebrate Paleontology/M. Ellison/© AMNH, 105 (bottom) Research Library/Image 845, 106 Department of Vertebrate Paleontology/M. Ellison/© AMNH, 107 Department of Vertebrate Paleontology/M. Ellison/© AMNH, 109 (top) Department of Vertebrate Paleontology/M. Ellison/© AMNH (bottom) Department of Vertebrate Paleontology/M. Ellison/© AMNH, 111 Research Library/Image 6744, 115 Department of Vertebrate Paleontology/M. Ellison/© AMNH, 116 Department of Vertebrate Paleontology/M. Ellison/© AMNH, 119 (top) Department of Vertebrate Paleontology/M. Ellison/© AMNH, (bottom) Department of Vertebrate Paleontology/Courtesy of Norell Lab, 122-123 Department of Vertebrate Paleontology/M. Ellison/© AMNH, 124 Department of Vertebrate Paleontology/M. Ellison/© AMNH, 132 Department of Vertebrate Paleontology/Archives, 133 Research Library/Image 19790, 135 Photo Studio/D. Finnin/© AMNH,137 Department of Vertebrate Paleontology Archives, 138 Research Library/Image 2A6933, 143 (top) Photo Studio/D. Finnin/© AMNH, (bottom left) © Sauriermuseum Aathal, Aathal, Switzerland, and Urs Möckli, moeckliurs@bluewin.ch., (bottom right) © AMNH, 144 (left) Research Library/Image 45615, 145 Research Library/Image 17506, p146-147 Photo Studio/M. Shanley/© AMNH, 149 (left) Department of Vertebrate Paleontology Archives, (right) Department of Vertebrate Paleontology Archives, 151 Research Library/Image 5409, 154 Photo Studio/D. Finnin/© AMNH, 159 (top) Office of the Registrar, 160 (top) Research Library/Image 7722, 161 Photo Studio/D. Finnin/© AMNH, 164-165 Photo Studio/D. Finnin/© AMNH, 166 (bottom) Photo Studio/R. Mickens/© AMNH, 167 Photo Studio/R. Mickens/© AMNH, 170 Research Library/Image 5414, 174 Research Library/Image 19508, 175 Research Library/Image 314804, 176 Research Library/Image 19449, 178 (bottom) Research Library/Image 5413, 183 (top) Research Library/Image 5783, (bottom) Photo Studio/D. Finnin/© AMNH, 184-185 Research Library/Image 7769, 186 Photo Studio/C. Chesek/© AMNH, 190-191 Department of Vertebrate Paleontology/M. Ellison/© AMNH, 192 Research Library/Image 5382, 193 (top) Photo Studio/D. Finnin/© AMNH, (bottom) Department of Vertebrate Paleontology, 198-199 Research Library/Image 3147, 200 Photo Studio/D. Finnin/© AMNH, 201 (top) Research Library/Image 5415, (bottom) Photo Studio/D. Finnin/© AMNH, 202-203 Research Library/Image 324091, 204 Photo Studio/D. Finnin/© AMNH, 205 (bottom) Research Library/Image 315066, 206-207 Photo Studio/D. Finnin/© AMNH, 210 Photo Studio/D. Finnin/© AMNH, 212-213 Research Library/Image 3773, 214 (top) Research Library/Image 201, (bottom) Department of Vertebrate Paleontology Archives, 215 (top) Department of Vertebrate Paleontology Archives, (bottom) Research Library/Image 330491, 216-217 Photo Studio/M. Shanley/© AMNH, 218-219 Photo Studio/D. Finnin/© AMNH, 219 (top) Department of Vertebrate Paleontology Archives, 220 Research Library/Image 707, 222 Research Library/Image 35876, 223 Research Library/Image 7740, 224 Research Library/Image 35044, 225 Department of Vertebrate Paleontology/M. Ellison/© AMNH, 227 Department of Vertebrate Paleontology/M. Ellison/© AMNH, 228 Photo Studio/M. Stanley/©AMNH, 229 Photo Studio/D. Finnin/© AMNH, 232 (top) Photo Studio/D. Finnin/© AMNH, 240 Research Library/Image 5419.

Agencies
4-5 Getty Images/Eric Van Den Brulle, 10 (top) Granger/REX/Shutterstock, (bottom) Heritage Auctions, HA.com, 11 (top) ILM (Industrial Light & Magic)/Amblin/Universal/Kobal/REX/Shutterstock, (bottom) Travellight, 12-13 John Orris/The New York Times/Re/eyevine, 18 (left) Chronicle/Alamy (right) Private Collection, 21 (top) Sergey Krasovskiy/Stocktrek Images/Getty Images, (bottom) Sean Murtha/www.seanmurthaart.com, 24 (top) Art Collection 2/Alamy, (bottom) Paul D. Stewart/Science Photo Library, 26 (left) Science History Images/Alamy, (right) Natural History Museum/Alamy, 27 (top) Chronicle/Alamy, (bottom) Natural History Museum/Alamy, 28 Topfoto/The Granger Collection, 29 Galyna Andrushko/Shutterstock, 32 Blickwinkel/Alamy, 38 John Weinstein/Field Museum Library/Getty Images, 41 Bob Elsdale/Getty Images, 46 Manchester University, 50 VANDERLEI ALMEIDA/AFP/Getty Images, 54-55 © Louie Psihoyos, 57 Natural History Museum, London/Science Photo Library, 60-61 © PNSO, 62 Visuals Unlimited, Inc/David Cobb, 63 (top) Jose Antonio Penas/Science Photo Library, (bottom) Kevin Schafer/Getty Images, 64-65 Rob Stothard/Getty Images, 67 (bottom) Jim West/Alamy, 68-69 © PNSO, 71 (top) John Downes/Getty Images, 72-73 © PNSO, 79 Bryan Smith/ZUMA Wire/Alamy, 80 Funkmonk (Michael B.H.), 81 Eduard Sola, 82-83 © PNSO, 84 (top) Matthias Kabel, 85 (top) The History Collection/Alamy, (bottom) Bidar et al, 87 ©PNSO, 89 (top) Julius T Csotonyi/Science Photo Library, (bottom) © Mohammad Haghani, 90-91 ©PNSO, 92 Bettmann/Getty Images, 93 (left) Granger Historical Picture Archive/Getty Images, 94 Sabena Jane Blackbird/Alamy, 95 (bottom) Dirk Wiersma/Science Photo Library, 98-99 ©PNSO, 100 © Mick Ellison, 102 (top) Robert m. Sullivan phd, 103 Xavier Fores - Joana Roncero/Alamy, 105 (top) © PNSO, 108 © Mick Ellison, 110 Didier Descouens, 112 (bottom) Bob Bakker, 113 © Mick Ellison, 114 Jonathan Blair/Getty Images, 117 Stocktrek Images, Inc./Alamy, 118 Martin Shields/Alamy, 120 O.Louis Mazzatenta/Getty Images, 121. VPC Travel Photo /Alamy, 125. © Mick Ellison, 126-127 © PNSO, 128 (left) Millard H.Sharp/Science Photo Library, (right) Benoitb/Getty Images, 129 Matteis/Look at Sciences/Science Photo Library, 130. Sabena Jane Blackbird/Alamy, 131 John Sibbick/Science Photo Library, 134 DeAgostini/UIG/Science Photo Library,136 The Natural History Museum/Alamy Stock Photo, 137 (top) © PNSO, 138-139 (top) © PNSO, (bottom) Ira Block/Getty Images, 140-141 Dan Kitwood/Getty Images, 142 Stocktrek Images, Inc./Alamy Stock Photo, 144 (right) Leon Werdinger/Alamy, 148 Archive PL/Alamy, 150 Steve Pridgeon/Alamy, 152-153 © PNSO, 154 (top) © Jason Brougham, 155 © Mick Ellison, 156-157 Wang Ping/Xinhua/Alamy, 158-159 © PNSO, 160 (bottom) Zack Frank/Shutterstock, 162-163 David Reed/Alamy, 166 (top) © PNSO, 168-169 © PNSO, 170-171 Justin Tallis/AFP/Getty Images, 171 (top) Science History Images/Alamy, 172-173 National Geographic, 177 Willem van Valkenburg from Delft, Netherlands Euoplocephalus Tyrrell, 178 (top) Andy Crawford/Dorling Kindersley/Getty Images, 178-179 Leonello Calvetti/Getty Images, 180-181 © PNSO, 182 Oleksiy Maksymenko/Getty Images, 187 Gilmanshin/Getty Images, 188-189 Millard H.Sharp, 189 Private Collection, 194-195 Richard T. Nowitz/Getty Images, 197 Smithsonian Institute/Science Photo Library, 198 (top) The Natural History Museum/Alamy, 199 (top) Universal History Archive/UIG via Getty Images, 205 (top) © PNSO, 207 Jos Dinkel, 208-209 © PNSO, 209 (top) Jim Page/North Carolina Museum of Natural Sciences/Science Photo Library, 209 (bottom) Corbin17/Alamy, 211 © PNSO, 216 (top) DeAgostini/UIG/Science Photo Library, 218 Photograph by Amy Martiny Heritage College of Osteopathic Medicine, 221 De Agostini/UIG/Science Photo Library, 226 Natural History Museum/Alamy, 230 Getty Images, 230-231 Mark Garlick/Science Photo Library, 232 NASA, 234 Amy Paturkar, 235 Alan Majchrowicz/Getty Images

Illustrations on pages 20, 33, 42 by Geoff Borin/Carlton Publishing Group

Every effort has been made to acknowledge correctly and contact the source and/or copyright holder of each picture and Carlton Publishing Group apologises for any unintentional errors or omissions, which will be corrected in future editions of this book.

ACKNOWLEDGEMENTS

I would like to acknowledge my colleagues, fellow dinosaur paleontologists both living and dead, and especially my students. They are the ones whose discoveries in the field and in the lab make dinosaur paleontology so vital and fun. To those I have shared campfires, bar stools and hotels with, you, my road companions, are a large part of this. The American Museum of Natural History (AMNH) is recognized for its generous support for science, collections, education and exhibitions. For supporting many of my efforts, acknowledgement goes out to the Macaulay family for their generous support of my position. Mick Ellison (Division of Vertebrate Paleontology) and Joanna Hostert (Global Business Development) organized the images.

The staff of the Division of Paleontology, the AMNH Research Library and the AMNH Photo Studio (particularly Director of the Photography Studio Denis Finnin), helped in locating pictures and specimens. The Global Business Development team (led by Sharon Stulberg) at AMNH, and the editorial staff at Carlton books (UK) are thanked. Without them this book would never have happened and if it had it would not be as good. Daniel Barta is thanked for a careful reading of this manuscript. For assistance in China, I thank Liu Jie, Xu Xing, Gao Ke-Qin and the people at PNSO, especially Yang Yang, Li Qing, and Zhou. Lastly Inga and Vivian are recognized for putting up with me through this.

INDEX